AF183798

Underutilized Plants Based-Nanoparticles: Herbal Remedy to Treat Human Diseases

Eugenie Nepovimova • Kamil Kuca
Lukas Peter • Nidhi
Editors

Underutilized Plants Based-Nanoparticles: Herbal Remedy to Treat Human Diseases

Underutilized Plants and Nanoparticles

Springer

Editors
Eugenie Nepovimova
Department of Chemistry
University of Hradec Kralove
Hradec Kralove, Czech Republic

Kamil Kuca
Center for Basic and Applied Research
Faculty of Informatics and Management
University of Hradec Kralove
Hradec Kralove, Czech Republic

Lukas Peter
Faculty of Materials Science
and Technology
Technical University of Ostrava
Ostrava, Czech Republic

Nidhi
Thakur P.G. College of Education, Dhaliara
Kangra, Himachal Pradesh, India

ISBN 978-3-032-20383-0 ISBN 978-3-032-20384-7 (eBook)
https://doi.org/10.1007/978-3-032-20384-7

© The Editor(s) (if applicable) and The Author(s), under exclusive license to Springer Nature Switzerland AG 2026

This work is subject to copyright. All rights are solely and exclusively licensed by the Publisher, whether the whole or part of the material is concerned, specifically the rights of translation, reprinting, reuse of illustrations, recitation, broadcasting, reproduction on microfilms or in any other physical way, and transmission or information storage and retrieval, electronic adaptation, computer software, or by similar or dissimilar methodology now known or hereafter developed.
The use of general descriptive names, registered names, trademarks, service marks, etc. in this publication does not imply, even in the absence of a specific statement, that such names are exempt from the relevant protective laws and regulations and therefore free for general use.
The publisher, the authors and the editors are safe to assume that the advice and information in this book are believed to be true and accurate at the date of publication. Neither the publisher nor the authors or the editors give a warranty, expressed or implied, with respect to the material contained herein or for any errors or omissions that may have been made. The publisher remains neutral with regard to jurisdictional claims in published maps and institutional affiliations.

This Springer imprint is published by the registered company Springer Nature Switzerland AG
The registered company address is: Gewerbestrasse 11, 6330 Cham, Switzerland

If disposing of this product, please recycle the paper.

Foreword

The rapid merging of plant sciences, nanotechnology, and medicine has opened transformative avenues for addressing some of the most pressing global health challenges. In this context, *Underutilized Plants Based-Nanoparticles: Herbal Remedy to Treat Human Diseases* arrives as a timely and much-needed contribution that bridges traditional herbal knowledge with cutting-edge nanoscience. The book highlights how underutilized plant biodiversity, which are often overlooked despite its immense therapeutic potential, can be harnessed sustainably to develop next-generation nanomaterials for human health applications.

This volume provides a comprehensive and well-structured exploration of biogenic nanoparticles derived from plants, emphasizing their role as eco-friendly, cost-effective, and biologically compatible alternatives to conventional synthetic nanomaterials. Beginning with the principles of plant-based nanoparticle synthesis, the book elucidates the unique physicochemical and biological properties that make these nanomaterials highly effective against a range of human health disorders. The detailed discussion of antibacterial mechanisms against both Gram-positive and Gram-negative bacteria, particularly those associated with food and waterborne infections, is especially valuable in an era of escalating antimicrobial resistance.

Equally compelling is the coverage of antifungal applications, where plant-derived nanomaterials are evaluated against pathogenic yeasts and fungi, supported by current experimental evidence. The book further extends into the realm of nanomedicine, presenting critical insights into the use of metal and metal-oxide nanoparticles for treating cancer, metabolic disorders, and other acute and chronic diseases. By integrating mechanistic pathways, in vitro and in vivo findings, and translational challenges, the authors provide a balanced and scientifically rigorous narrative.

Overall, this book serves as an important reference for researchers, clinicians, policymakers, and students interested in sustainable nanomedicine and the responsible utilization of underexplored plant resources. It underscores the promise of underutilized plant-based nanoparticles in advancing global healthcare while promoting environmental sustainability and biodiversity conservation.

I strongly recommend this book to researchers, academicians, and students in the fields of nanotechnology, pharmacology, biotechnology, and herbal and traditional medicine. It will serve both as an authoritative reference and as an inspirational resource, fostering innovative research and advancing knowledge in the rapidly evolving field of plant-based nanotherapeutics.

Soumya Ghosh
Assistant Professor
Microbiology Department
Natural and Medical Sciences Research Center
University of Nizwa
Nizwa, Oman

Preface

The utilization of plants plays a pivotal role in ensuring global health, survival, and safety, particularly in addressing the adverse effects of chemical agents commonly consumed through food, beverages, agriculture, and pharmaceuticals. To achieve sustainability and combat human illnesses, scientific advancements increasingly focus on strategies involving plants, especially underutilized ones. These underutilized plants hold immense potential in addressing health issues due to their diverse phytochemical profiles, which include compounds such as phenols, alkaloids, terpenoids, saponins, polysaccharides, anthraquinones, and phytosterols. These phytochemicals contribute to their significant therapeutic properties, including antioxidant, anti-inflammatory, antidiabetic, anti-Alzheimer's, antiepileptic, antimicrobial, antidiuretic, hepatoprotective, and anticancer activities. These properties not only reduce the burden of synthetic chemicals on the human body but also lower treatment costs for society. Despite their proven efficacy in traditional medicine, underutilized plants have been largely neglected in modern healthcare systems, exacerbating health issues such as malnutrition, food insecurity, and the rising costs of medical treatments. However, the use of underutilized plants for treatment does pose challenges. These include insufficient dosing through intravenous administration, reduced biological activity under physiological conditions, and the potential loss of efficacy during isolation. Addressing these limitations requires innovative approaches.

Nanotechnology is an emerging area that could serve as a vehicle for the future development of products and processes that will resolve some of the issues surrounding traditional medicine (conventional). It has the potential to improve the effectiveness and usability of plant-based therapeutic products by harnessing the bioactive elements contained within the plants themselves. The physical properties of nanotechnology, most notably the ability of nanotechnology products to be manufactured easily at small sizes, along with the lesser toxicity of these nanosized products improves the ability of your body to absorb and utilize the therapeutic benefits of the plant-based product. It has been demonstrated that nanosized products can penetrate difficult biological barriers (e.g., BBB). Therefore, by using nanotechnology in conjunction with underutilized plants to create new therapeutic products, we may have the potential to create highly focused and less expensive product solutions for a variety of medical conditions, including but not limited to, infections caused by microbes (bacteria, fungi, and yeast); acute medical conditions; and chronic medical conditions.

The target audience of this publication includes researchers, academics, clinicians, and postgraduate students in the areas of Nanotechnology, Pharmaceutical Sciences, Biotechnology, Herbology, and Biomedical Research. This book serves as a scientific bridge between contemporary methods and techniques in Nanosciences and the Ancient traditions of Medicinal Knowledge. This book promotes cross-discipline innovation and the creation of plant-based nanotherapeutics that are sustainable and promote health.

The editors, authors, and co-authors express heartfelt thanks and appreciation for the invaluable support provided by Dr. Soumya Ghosh and Dr. Kanika Dulta through their guidance, constructive feedback, and expertise that has enhanced the scientific quality and direction of this book.

Hradec Kralove, Czech Republic — Eugenie Nepovimova
Hradec Kralove, Czech Republic — Kamil Kuca
Ostrava, Czech Republic — Lukas Peter
Kangra, Himachal Pradesh, India — Nidhi

About the Book

This book covers a wide array of biogenic nanoparticle applications. It focuses on the various applications of underutilized plant-based nanomaterials as sustainable herbal alternatives for medicinal use in treating human health and diseases.

The first section details how plant-based nanomaterials can be utilized to produce nanomaterials, as well as describes their unique biological properties and biological activity against human health-related issues. Next, it provides a comprehensive overview of how plant-derived nanomaterials function as antibacterial agents by outlining the mechanisms of antibacterial action and the resulting experimental evidence from recent research. It specifically focuses on the use of plant-derived nanomaterials against bacteria that are associated with food and water sources, and Gram-negative and Gram-positive bacteria and their associated mechanisms of action. It also describes how nanomaterials derived from plants can be used to combat fungal infections by providing an overview of the current antifungal activities of plant-derived nanomaterials, specifically by listing antifungal activity and common pathogenic yeast and fungal species. Together, these two sections of the book provide an extensive compilation of information regarding the opportunities for developing and utilizing plant-derived nanomaterials, and the role they play in addressing various health conditions through environmentally friendly means. The book expands upon the experiential usage of underused plant-derived nanoparticles to treat diseases including cancer and acute and chronic diseases and details how metal and metal oxide-based NPs help treat breast and prostate cancers, along with metabolic disorders like diabetes and obesity. It critically discusses mechanistic pathways, in vitro/vivo data, and translational issues. Moreover, it describes material for sustainable nanomedicine to support use of underutilized plant biodiversity for Global Healthcare practices.

Contents

About the Editors

Eugenie Nepovimova is currently serving as Assistant Professor in chemistry department at University of Hradec Kralove. Her research is on drug design and development focused on antiaging drugs, novel potential therapeutics against neurodegenerative diseases, and potential prophylactics and antidotes in case of organophosphate intoxication. She has served as guest editor in international journals. She was principal investigator of GA23-05857S (2023–2025, Czech Science Foundation), GA20-22037S (2020–2022, Czech Science Foundation), and Project Excellence of University of Hradec Kralove. She is also research team member of GA15-16701S, GA16-08554S, GA17-08596S, and GA18-01734S (all Czech Science Foundation), NV15-30954A and NV19-09-00578 (both Ministry of Health), and CZ.02.1.01/0.0/0.0/16_017/0002311 and CZ.02.1.01/0.0/0.0/16_025/0007444 (both OPVVV).

Number of citations (total: 7432/without self-citations: 7011)

H-index: 47

Kamil Kuca is a distinguished scholar who has held professorial positions and served as an invited or adjunct faculty member at esteemed academic institutions worldwide. His affiliations include the University of Defense in Brno, Czech Republic; Nanjing Agricultural University in Nanjing, China; University Hospital Hradec Kralove in Hradec Kralove, Czech Republic; Florida International University in Miami, USA; University of Hradec Kralove in Hradec Kralove, Czech Republic; Institute of Military Engineering in Rio de

Janeiro, Brazil; Yangtze University in Jingzhou, China; University Technical Malaysia in Kuala Lumpur, Malaysia; Technical University Kosice in Kosice, Slovakia; and University of Granada in Granada, Spain.

His research encompasses diverse areas within pharmacology and toxicology, drug design and discovery, medical device development, pharmacoeconomics, food science, technology transfer, chemical and biological threat assessment, and military research. He has served as a principal investigator or team member on numerous national and international projects funded by prestigious organizations such as Czech Science Foundation, Ministry of Health of the Czech Republic, Ministry of Education of the Czech Republic, European Union, and NATO. With an extensive network of collaborators spanning across the globe, including Korea, India, Croatia, the United Arab Emirates, China, Spain, Malaysia, and Brazil, he has contributed significantly to advancing scientific knowledge and fostering international cooperation.

Recognized for his scholarly impact, he boasts an impressive H-index of 70 and has accumulated over 20,000 citations in the Web of Science. In 2021, he was honored as a Highly Cited Researcher by Clarivate Analytics. Additionally, he holds five patents and serves as a scientific advisor to several technology companies, further demonstrating his leadership and expertise in the field.

Lukas Peter serves as Deputy Head of the Faculty of Material Science and Technology. He is also heading the position of Vise-Dean for cooperation with industry at the Technical University of Ostrava in the Czech Republic. His expertise is in designing and development of medical hardware and software like monitoring of cardiovascular systems, biomechanics of cardiovascular systems, signal processing in biomedicine, etc. Currently, under his supervision, PhD students are working on the design and creation of a predictive model for evaluating the risk of treatment in chronic lymphocytic leukemia, and design and development of a platform for reading and setting pacemaker firmware for a pacemaker tester. Presently, he is working on a project titled "Prototype - Innovative tester for automated testing of a dual-cavity external pacemaker" granted by the Ministry of Industry and Trade of the Czech Republic.

Nidhi (PhD, Botany) formerly served as Assistant Professor, Thakur P.G. College of Education, Dhaliara, Kangra, 177103, India. Currently, she is serving as postdoctoral fellow in the Center of Advanced Innovation Technologies, VŠB-Technical University of Ostrava, 70800, Ostrava-Poruba, Czech Republic. Her research involves optimizing synthesis process of metal oxides, formulation of doped nanomaterials of metal oxides and their characterization and evaluation of their biomedical potential. Furthermore, she is also skilled in exploring the secondary metabolites and the biological potential of underutilized plants. Her dedication to research is evident through her contributions to various prestigious journals, where she has published five insightful review articles, five research articles, three book chapters, and one edited book that enrich the fields of plant physiology, nanotechnology, and ethnopharmacology. Her multifaceted achievements and scholarly endeavors emphasize her commitment to academic excellence and her invaluable contributions to the field of botany and nanotechnology.

Abbreviations

ADT	Androgen Deprivation Therapy
AgNPs	Silver Nanoparticles
AgNPs	Silver Nanoparticles
Akt/PI3K	Protein Kinase B/Phosphatidylinositol 3-Kinase
AMR	Antimicrobial Resistance
AR	Androgen Receptors
ATM	Ataxia-Telangiectasia Mutated
AuNPs	Gold Nanoparticles
BMI	Body Mass Index
BRCA1	Breast Cancer gene 1
BRCA2	Breast Cancer Gene 2
CaONPs	Calcium Oxide Nanoparticles
CAT	Catalase
CFU	Colony Forming Units
CuNPs	Copper Nanoparticles
DLS	Dynamic Light Scattering
DNA	Deoxyribonucleic Acid
DPPH	2,2-Diphenyl-1-Picrylhydrazyl
E. coli	*Escherichia coli*
EC50	Half-maximal effective concentration
EDS	Energy Dispersive X-Ray Spectroscopy
FAO	Food and Agriculture Organization
FCC	Face-Centered Cubic
FTIR	Fourier Transform Infrared Spectroscopy
GPX	Glutathione Peroxidase
GRAS	Generally Recognized as Safe
IL-1β	Interleukin-1 Beta
IL-6	Interleukin-6
MAPK/ERK	Mitogen-Activated Protein Kinase/Extracellular Signal-Regulated Kinase
MC4R	Melanocortin 4 Receptor Gene

MDA	Malondialdehyde
MFC	Minimum Fungicidal Concentration
MO-NPs	Metal and Metal-Oxide Nanoparticles
MRP	Multidrug Resistance Associated Protein
MRSA	Methicillin-Resistant *Staphylococcus aureus*
NAFLD	Non-Alcoholic Fatty Liver Disease
NF-κB	Necrosis Factor Kappa B
NPs	Nanoparticles
NUS	Neglected and Underutilized Species
OC	Oligo-Chitosan
PCOS	Polycystic Ovarian Syndrome
PCSK1	Proprotein Convertase Subtilisin/Kexin Type 1
pH	Potential of Hydrogen
POMC	Pro-opiomelanocortin Gene
PSA	Prostate-Specific Antigen
QS	Quorum Sensing
ROS	Reactive Oxygen Species
SEM	Scanning Electron Microscopy
SOD	Superoxide Dismutase
SPR	Surface Plasmon Resonance
STAT	Signal Transducer and Activator of Transcription
STZ	Streptozotocin
TEM	Transmission Electron Microscopy
TNF-α	Tumor Necrosis Factor-Alpha
TSCA	Toxic Substances Control Act
WHO	World Health Organization
XRD	X-Ray Diffraction
ZnONPs	Zinc Oxide Nanoparticles

List of Figures

List of Tables

Chapter 1
Plant-Based Nanotechnology: Underutilized Plants as a Source of Green Nanoparticles

Priyanka Sharma and Mamta Sharma

Abstract The term underutilized plants refers to categories of plant species whose potential remains largely untapped. In recent times, these plants have gained recognition due to their green synthesis of nanoparticles. Species such as *Moringa oleifera*, *Kaempferia angustifolia*, *Tephrosia purpurea*, *Nasturtium officinale*, *Clerodendrum*, and *Calotropis* have been utilized because they contain significant antimicrobial, anticancer, and antioxidant properties. Secondary metabolites, including terpenes, phenolics, and organic acids, are mainly produced as a defense mechanism, which further helps to fabricate nanoparticles by functioning as reducing and stabilizing agents. Traditional nanoparticle synthesis methods often involve the use of harmful substances. On the other hand, the green synthesis of phyto nanoparticles utilizing plant extracts and phytochemicals introduces a sustainable and eco-friendly substitute. This chapter highlights the ability of underutilized plants to synthesize nanoparticles and emphasizes their phytochemical diversity and therapeutic implications.

Keywords Underutilized plants · Green synthesis · Phytochemicals · Nanoparticles

Abbreviations

AgNPs	Silver nanoparticles
AuNPs	Gold nanoparticles
DLS	Dynamic light scattering
EDS	Energy dispersive X-ray spectroscopy
FTIR	Fourier transform infrared spectroscopy
NPs	Nanoparticles

P. Sharma (✉) · M. Sharma
Department of Botany, Sardar Patel University, Mandi, Himachal Pradesh, India

© The Author(s), under exclusive license to Springer Nature Switzerland AG 2026

E. Nepovimova et al. (eds.), *Underutilized Plants Based-Nanoparticles: Herbal Remedy to Treat Human Diseases*, https://doi.org/10.1007/978-3-032-20384-7_1

SEM	Scanning electron microscopy
TEM	Transmission electron microscopy
XRD	X-ray diffraction
ZnONPs	Zinc oxide nanoparticles

1.1 Introduction

Nanotechnology studies materials at the nanoscale in general between 1 and 100 nm. It works at this extremely small scale to provide various benefits across diverse scientific disciplines like bioengineering, dentistry, and pharmaceutical formulations [1]. Over the years due to their extensive use in photonics, catalysis, electronics, medicines, and sensing, NPs formation has gained significant level of attention. NPs can be synthesized by utilizing physical, chemical, and green (biological) approaches [2]. The chemical and physical methods may generate harmful byproducts and are often costly. Physical methods such as diffusion, laser ablation, electrolysis, evaporation condensation, sputter deposition, pyrolysis, and high-energy ball milling are commonly used to synthesize nanoparticles (NPs). Chemical methods of synthesis involve thermal decomposition, chemical reduction, electrochemical, and micro-emulsion. These methods have many limitations like low production rate, high energy consumption, and environmental degradation due to toxic byproduct formation [3].

In contrast, biological methods are easy to synthesize, cost-effective, and reduce chemical waste during production. Nanoparticles (NPs) synthesized from biological materials are known as green-synthesized NPs [4]. This strategy avoids the use of harmful chemicals, consumes less energy, and produces environmentally friendly byproducts [5]. Plant extract and microorganisms are utilized to form NPs in biological methods [6]. Plants contain key phytochemicals like alkaloids, glycosides, flavonoids, polyphenols, and terpenoids, etc., in higher concentrations compared to other natural sources, algae, bacteria, and fungi. These phytoconstituents serve as a capping, reducing, and stabilizing agent to synthesize green NPs. These bioactive compounds not only improve the biological properties of NPs but also reduce health and environmental risks [7]. Various biological activities such as anticancer, anti-inflammatory, antidiabetic, and antioxidant activities are present in underutilized plants due to diverse and potent phytochemical profiles; such plants are gaining recognition as a notable source for NPs synthesis [8]. Different parts of plants like bark, leaves, stems, roots, peel, flowers, fruit, and seeds are utilized to synthesize nanoparticles [9]. Therefore, green synthesis is considered a safer strategy to synthesize nanoparticles as it mandates comparatively low energy for reaction initiation and is eco-friendly [10].

1.2 Fundamentals of Green Nanotechnology

Underutilized plants are those that were traditionally used for multiple purposes, such as food, fodder, and medicines, but their wider potential remains mostly untapped. Terms such as neglected, minor, underdeveloped, underexploited, lost, orphan, promising, and alternative plant species are often used to describe such plants [11]. Such plants offer important benefits including boosting nutrition, diet diversity, health, and food availability. Still many plants in nature remain underexplored for different reasons [12]. These plants are rich in bioactive compounds, help preserve biodiversity, offer health benefits, and promote local economies. Despite these advantages challenges like limited awareness, lack of proper approach, insufficient national schemes for reintroduction, inadequate wide-scale cultivation, and low utilization of such plants have hindered their widespread use [13]. Addressing these issues is crucial as efficient utilization of these neglected natural resources could expand the availability of herbal medicines and plant-based drugs, so improving the welfare of the present and future generations. So there is an immediate need to recognize the significance of these underutilized plants and safeguard their genetic diversity to ensure nutritional and therapeutic supply for future [14].

For example, the fruit of *Elaeagnus pyriformis* is considered underutilized and it has been known to synthesize silver nanoparticles. The plant is rich in phytochemicals including long-chain fatty acids and their derivatives which act as reducing and stabilizing agents at particular temperature and conditions [15]. The mechanism for the synthesis of NPs includes reduction and stabilization by phytochemicals present in plant extract which leads to the nucleation and growth of nanoparticles. The resulting NPs are then characterized to confirm their physical, chemical, and functional attributes [2] (Fig. 1.1).

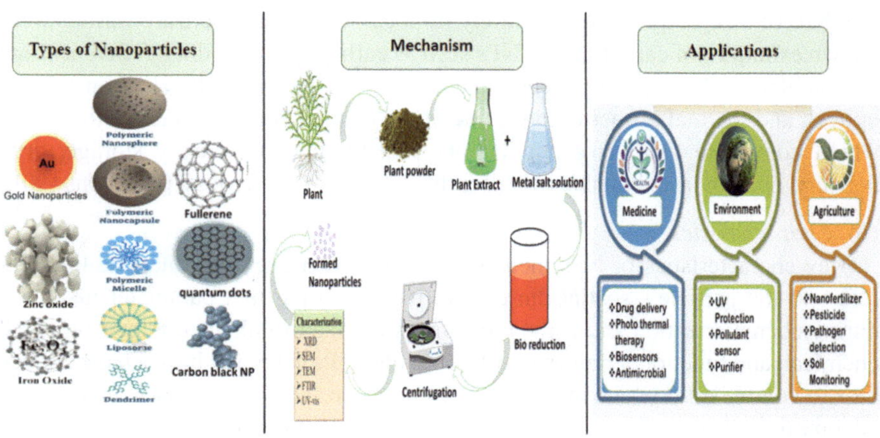

Fig. 1.1 Types, mechanisms, and applications of NPs

1.3 Classification, Characteristics, and Methods of Characterization of NPs

(a) *Classification*

Organic NPs
These NPs are composed of natural or synthetic organic molecules with a diameter between 10 and 100 nm or less. They are composed of carbohydrates, lipids, and protein, and exhibit traits including biodegradability and nontoxicity. Such properties make them highly suitable in diverse therapeutic areas like cancer therapy, targeted drug delivery, and bioimaging, etc. Therapeutic efficiency of organic NPs is determined by shape, surface morphology, and size [16]. Common examples of organic NPs (as shown in Fig. 1.1) include dendrimers, micelles, nanogels, liposomes, layered biopolymers, and polymeric NPs [17].

Inorganic NPs
These NPs lack carbon atoms and are usually made from metals or metal oxides. Over the last few decades, metal oxides have gained the attention of researchers because of their remarkable qualities. Metal oxidebBased NPs are made by utilizing ionic interactions where positively charged metal ions and negatively charged oxygen ions combine to form metal oxides. Aluminum oxide (Al_2O_3), titanium oxide (TiO_2), silicon dioxide (SiO_2), and zinc oxide (ZnO) are the most commonly studied examples [18] (Fig. 1.1). Metal NPs are composed entirely of metals such as Cu, Ag, and Au NPs. These are characterized by their strong localized surface plasmon resonance (LSPR) which gives them distinct electrical characteristics. Scientist often tweak these NPs by adjusting their shape, size and surface properties to get the exact performance they want [19].

Carbon-Based NPs
The NPs of this group are made entirely from carbon atoms. There are various well-known examples of carbon-based NPs such as carbon quantum dots, fullerenes, and carbon black NPs (Fig. 1.1). Electrical conductivity, high strength, electron affinity, optical, and thermal properties are some of the distinct characteristics of these NPs which contribute to diverse applications like bioimaging, energy storage, drug delivery, and environmental sensing for the detection of microbial pathogens [20].

(b) *Characteristics*

Key characteristics of NPs include size, shape, surface area, and stability. *Size* refers to the particle's spatial dimension. It is often the first parameter measured during characterization and is the most important factor influencing their physical, chemical, and functional properties [21]. Smaller NPs generally have a large surface area to volume ratio which makes them more stable. The large surface area of such nanoparticles leads to higher interaction [22]. NPs are mostly spherical in shape but they can also appear as ellipsoids, needles, or cuboids. NPs can be organized into many structural arrangements like isolated particles, chains, or flocks due to their complex internal structure. Morphology and organization of NPs are important in the determination of functional and physicochemical properties [23]. The stability of NPs depends on their tendency to aggregate or react with the surrounding media. This can be assessed through surface complexation which regulates colloidal stability [10].

(c) *Methods for Characterization*

UV-Visible Spectroscopy Light is used by UV visible spectroscopy to analyze the optical properties of NPs which is a simple and widely used technique. For example, a key characteristic band for Ag NPs is typically displayed near 436 nm. Therefore, it helps to identify nanoparticles based on optical properties [24].

FTIR In this technique, material is irradiated with infrared light and transmitted or absorbed radiation is recorded. The resulting spectrum gives information about bonds, polarity, and oxidation state of the sample. This characterization technique is largely utilized to determine the surface chemical composition of NPs [20].

XRD This technique analyzes the diffraction pattern which identifies the structure of the material in crystalline samples. It gives information about size, how atoms are arranged, and the shape of the crystal [25].

SEM The SEM technique uses electrons to look at nanoparticles which is based on the electron scanning principle. It produces high-resolution image providing all available information of NPs [26].

TEM Its core idea is built on the transmittance, so it confers 2D details about the NPs. It also uses electrons and provides even higher magnification than SEM, making it possible to get details about the internal structure and layers of NPs. For instance, using TEM, the quadrupolar hollow shell structure of Co_3O_4 NPs has been observed [26].

DLS It is used to find hydrodynamic particle size and distribution across a range of sizes. It measures the light interference produced by the Brownian motion of NPs. The results are shown by using the polydispersity index (PDI) which tells about the particle size distribution range, with its values ranging from 0 to 1. If the PDI value is close to 1, they are highly heterogeneous. If the PDI value is near to 0, the particles are mostly of the same size. While sample preparation should be done with care, slight contamination can lead to inappropriate results. This technique also works well for nonspherical particles like rods, using multiangle DLS [27].

1.4 Underutilized Plants and Their Integration with Nanotechnology

Underutilized plants are those plants that have lots of potential but have not been fully explored yet. Although these plants are loaded with nutrients and have beneficial health outcomes, they can be important for food, medicine, and even the replacement of chemicals, but there is still a lot more which is not been studied yet [28]. Researchers are slowly focusing more on underutilized plants due to their ability to support the environment and lower the risk associated with agricultural production. Many of these plants produce compounds that act as an efficient source of carbohydrates, essential fatty acids, minerals, phytochemicals, proteins, and vitamins. They also show anticancer, anti-inflammatory, and antioxidant properties which enhance the importance of these crops [13].

A few examples of underutilized plants used in nanoparticle synthesis are the following:

Moringa oleifera is known for its rapid growth and ability to stand with tough environmental conditions. It is used to synthesize a wide variety of metal and metal oxide NPs such as silver, gold, titanium oxide, iron oxide, and ZnONPs. This is mainly due to the occurrence of diverse secondary metabolites that enable NPs formation by acting as the reducing, stabilizing, and capping agents [29]. For instance, a change in color from pale yellow to brown which is due to surface plasmon resonance confirms the AgNPs synthesis from this plant. The semicubic crystal structure of these AgNPs was revealed by XRD. FTIR analysis displayed absorption bands corresponding to carbonyl, hydroxyl, and amine groups indicating the involvement of proteins and polyphenols from the *Moringa* extract. *TEM* revealed predominantly spherical NPs with a typical size of *50–60 nm* while *SEM* images confirmed uniform morphology and slight aggregation [30] (as Table 1.1 shows). Different plant parts such as leaves, flowers, seeds, and stems are leveraged to make NPs. However, the leaf extract of *M. oleifera* is most commonly utilized as it is easily available and serves as a rich source of metabolites [31].

Kaempferia angustifolia is an ethnomedicinally important aromatic herb of the family Zingiberaceae. The rhizome extract of this plant was used to synthesize AgNPs through the green synthesis method. The resulting AgNPs exhibited a characteristic UV Vis absorption peak at ~420 nm. Diverse particle shapes such as oval, rod, trigonal, hexagonal, and spherical of size ranging between 10 and 60 nm in diameter were observed through the TEM analysis. Crude rhizome extract and AgNPs of this underutilized medicinal plant exhibited antimicrobial activities against multidrug-resistant human urinary tract-infecting pathogenic strains [32] (see Table 1.1).

Tephrosia purpurea, commonly known as sharpunkha, is a wild herb that grows in dry, rocky, and sandy soil. The plant possesses essential activities like wound healing, antimicrobial, antileishmanial, and anti-inflammatory properties. It is rich in diverse phytochemicals which help in the synthesis of biogenic metal NPs [33]. The extract of this plant can be used to synthesize both silver and gold

Table 1.1 Underutilized plant-derived nanoparticles: synthesis, characterization, and biological potential

Sr. No.	Underutilized plant species (part used)	NPs synthesized	Characterization techniques	Biological activity	Shape, size	References
1	*Moringa oleifera* (leaf extract)	Ag	UV–Vis, XRD, TEM, SEM/EDX, FTIR	Antibacterial antioxidant	Spherical, ~50–60 nm	[30]
2	*Tephrosia purpurea* (leaf)	Ag	UV–Vis, XRD, FTIR, FE-SEM	Antibacterial, anticancer	Spherical, ~20–120 nm	[35]
3	*Calotropis gigantea* (latex/leaf)	ZnO	UV–Vis, XRD, FTIR, TEM	Antibacterial, antioxidant	Hexagonal/spherical, ~21.8 nm	[41]
4	*Clerodendrum inerme* (leaf)	Au	UV–Vis, XRD, TEM, FTIR	Antibacterial, Antioxidant, Antimycotic	Spherical, ~5.82 nm	[37]
5	*Kaempferia angustifolia* (rhizome)	Ag	UV–Vis, TEM/HRTEM, DLS, EDX	Antibacterial (UTI pathogens)	Mixed morphology (spherical/triangular), 10–60 nm	[32]
6	*Vitex negundo* (leaf)	Ag	UV–Vis, XRD, DLS, SEM/EDX, FTIR	Antibacterial, Anticancer	Spherical, 90–120 nm	[42]
7	*Nasturtium officinale* (watercress leaf)	Au	UV–Vis, EDS, TEM, FTIR	Anticancer, antioxidant	Spherical, 56.4 nm	[43]
8	*Trianthema portulacastrum* (leaf)	ZnO	UV–Vis, XRD, SEM, FTIR	Anti-inflammatory, Anticancer, Antioxidant	Spherical, >30 nm	[44]
9	*Oxalis corniculata* (leaf)	CuO	UV–Vis, TEM, FTIR	Antibacterial, antidiabetic potential	Spherical, 20–36 nm	[44]

nanoparticles. In one study, AuNPs were synthesized by using leaf extracts, which potentially inhibit the growth of *Escherichia coli*, *E. faecalis*, *K. pneumonia*, and *S. aureus* [34]. AgNPs were also synthesized with water-based leaf extract of the plant *T. purpurea*. The formation was confirmed by UV–Vis characterization showing a maximum absorbance at 436 nm. SEM data showed its spherical shape with an average size of ~100 nm. XRD showed diffraction peaks characteristic of crystalline silver, and FTIR detected the presence of alcoholic as well as aromatic groups [35] (as Table 1.1 shows).

Clerodendrum: Plants in this genus have many bioactive phytochemicals like phenylpropanoids, diterpenoids, and steroids, which give them significant anticancer and other cytotoxic properties. Certain species such *as C. inerme*, *C. viscosum*, *C. infortunatum*, *C. phlomidis*, and *C. glandulosum* are known to synthesize AgNPs and possess antimicrobial, antioxidant, cytotoxic, and antiparasitic activity. Aluminum and zinc oxide nanoparticles are synthesized from *C. phlomidis*, *C. heterophyllum*, and *C. infortunatum*. For gold nanoparticles, only *C. infortunatum* and *C. inerme* have been studied for their antioxidant, antibacterial, and antiproliferative [36]. XRD, UV-Visible, FTIR, TEM, EDX, and DLS were used to characterize green-synthesized NPs. TEM images showed AuNPs synthesized from *C. inerme* leaves extract are spherical in morphology; FTIR spectra indicated the occurrence of several functional groups bound to the NPs surface, acting as reducing and capping agents [37].

Calotropis is a perennial shrub of the family Asclepiadaceae and has long been used in traditional medicine. Among its species, *Calotropis gigantea* and *C. procera* are particularly important; they exhibit diverse pharmacological activities such as anticarcinogenic and antifertility effects, and their parts are also used in the treatment of skin infections [38]. A study revealed that the flower extract of *C. gigantea* was able to synthesize silver and gold NPs, where the extract acted as both a reducing and stabilizing agent [39]. Scientists have used plant extracts of *C. procera* to synthesize ZnONP, AgNPs, and AuNPs [40]. For example, ZnO NPs were synthesized from the latex of *C. gigantea*, where several characterization techniques were used. UV-Vis spectrum showed a strong absorption peak at 346 nm, therefore ensuring the presence of ZnO. FTIR spectra also revealed peaks at 493, 555, 601, 694, and 879 cm^{-1}, which show the presence of metal oxide groups, indicating the formation of ZnO NPs. The hexagonal crystal system was confirmed by XRD. The ZnONP size range according to the SEM images was about 50–200 nm [41].

1.5 Biological Activities and Applications of Underutilized Plant-Based Nanoparticles

Metallic nanoparticles are proven to be really effective in fighting against bacteria [45]. These nanoparticles increase stress in the bacterial cell by simultaneously targeting many biomolecules in bacteria. Their tiny particle size, large surface area, and good mechanical stability enable their use in medicine, specifically in antibacterial treatments. Plant-based NPs also offer eco-friendly alternatives due to the presence of different chemicals which enhance both stability and activity of NPs [46]. For example, AgNPs synthesized from *Vitex negundo* have shown strong antibacterial activity against Gram-positive and Gram-negative bacteria. Particularly, the Gram-positive strains such as *M. luteus* and *B. subtilis* have shown higher susceptibility relative to Gram-negative bacteria such as *E. coli* and *Salmonella typhi* [42]. Anticancerous potential is another major biological activity of plant-based NPs these metal or metal oxide NPs have shown potential in controlling the growth of cancer cells, including HeLa, HepG2, and HCT 116 cell lines. They help stop cancer by neutralizing harmful free radicals which can cause disease by damaging cells [47]. In one study, AuNPs synthesized from *Nasturtium officinale* were found to be efficient particularly for lung cancer treatment [43]. One major activity of plant-based NPs is their anti-inflammatory properties which help to reduce inflammation occurring due to diseases and injuries. Additionally, cells are protected from oxidative damage by scavenging free radicals derived from nanoparticles which are obtained from antioxidant rich plants [48]. For instance, ZnONPs synthesized from *Trianthema portulacastrum* (commonly known as Horse Purslane) are recognized for their notable anti-inflammatory properties [44].

In the field of medicine, these NPs are getting the spotlight especially for diagnostic and targeted drug delivery because of their biocompatibility and biodegradability. Polyphenols, flavonoids, and other phytochemicals act as a source to synthesize nanoparticles; they reduce toxicity and improve treatment efficiency [49]. Due to their nontoxic and biodegradable nature, plant-derived NPs are good for use in medical products, bandages, and contrast agents for imaging [50]. In agriculture, plant-based nanoparticles are used to promote plant growth, improve soil health, and reduce soil-borne pests and pathogens. Compared to traditional methods, plant-based nanoparticles are eco-friendly, cost-effective, and biocompatible for crop management and productivity enhancement [51] (Fig. 1.1).

1.6 Challenges and Limitations

1.6.1 Lack of Standardization in Synthesis

Using green synthesis methods to synthesize nanoparticles has substantial potential especially for medicines and environmental cleanup. However, the lack of standardization and reproducibility is the major challenge. Things like the type of plant, how the plant grows, and extraction methods can lead to inconsistent NPs quality and performance which makes reproducibility difficult [52].

1.6.2 Limited Awareness and Accessibility of Underutilized Plants

Many underutilized plants haven't been widely used because people don't know much about them and are often neglected due to agronomic, genetic, economic, social, and cultural reasons. These underutilized plants lack commercial value; most farming focuses on crops like rice, wheat, and maize, thereby leaving other helpful plants forgotten. As a result, they remain under researched mostly because of limited awareness [13].

1.6.3 Regulatory and Ethical Issues

As the science of nanoparticles advances, ethical and regulatory concerns have emerged regarding the use of NPs synthesized by plants. When bulk material is transformed into nanoscale form, it can behave very differently as the obtained nanomaterial can change significantly due to distinct reactivity, both in a lab setting and in living organisms. Even though using biological methods to make NPs has several benefits, proper regulation and risk assessment are still at an early stage [53].

1.7 Conclusion

Plant-based NPs have gained notable attention as a result of environmentally safe characteristics, biocompatibility, and economically viable production. They are rich in phytochemicals which contribute to several activities like antibacterial, anticancerous, and anti-inflammatory properties. This enables their application in medicine for targeted drug delivery and in agriculture for acting as bio-fertilizers. Despite their potential, several challenges remain such as a lack of standardization in synthesis, limited awareness and accessibility of such underutilized plants, and regulatory and ethical issues related to their safety and impact. Overcoming these limitations through interdisciplinary research, awareness about such plants, and clear regulatory frameworks can set the stage for plant-based nanoparticles in modern science.

Acknowledgments The author gratefully acknowledges the support and guidance of Dr. Mamta Sharma, Assistant Professor, Sardar Patel University, Mandi, and Dr. Nidhi Bhardwaj, Post-Doc Position at VSB, Technical University of Ostrava, for their valuable inputs while preparing this chapter.

Author Contribution PS—Writing original draft, data collection, conceptualization; MS—outline suggestion, review, and final approval.

Competing Interest The author declares no competing interests.

Ethical Approval Not applicable.

References

1. Gour A, Jain NK (2019) Advances in green synthesis of nanoparticles. Artif Cells Nanomed Biotechnol 47(1):844–851. https://doi.org/10.1080/21691401.2019.1577878
2. Khan F, Shariq M, Asif M et al (2022) Green nanotechnology: plant-mediated nanoparticle synthesis and application. Nanomaterials 12(4):673. https://doi.org/10.3390/nano12040673
3. Dikshit PK, Kumar J, Das AK et al (2021) Green synthesis of metallic nanoparticles: applications and limitations. Catalysts 11(8):902. https://doi.org/10.3390/catal11080902
4. Rana A, Yadav K, Jagadevan S (2020) A comprehensive review on green synthesis of nature-inspired metal nanoparticles: mechanism, application, and toxicity. J Clean Prod 272:122880. https://doi.org/10.1016/j.jclepro.2020.122880
5. Kanwar R, Rathee J, Salunke DB et al (2019) Green nanotechnology-driven drug delivery assemblies. ACS Omega 4(5):8804–8815. https://doi.org/10.1021/acsomega.9b00304
6. Ijaz I, Gilani E, Nazir A et al (2020) Detail review on chemical, physical and green synthesis, classification, characterizations and applications of nanoparticles. Green Chem Lett Rev 13(3):223–245. https://doi.org/10.1080/17518253.2020.1802517
7. Dubey S, Virmani T, Yadav SK et al (2024) Breaking barriers in eco-friendly synthesis of plant-mediated metal/metal oxide/bimetallic nanoparticles: antibacterial, anticancer, mechanism elucidation, and versatile utilizations. J Nanomater 2024(1):9914079. https://doi.org/10.1155/2024/9914079
8. de Araujo FF, de Paulo Farias D, Neri-Numa IA et al (2021) Underutilized plants of the Cactaceae family: nutritional aspects and technological applications. Food Chem 362:130196. https://doi.org/10.1016/j.foodchem.2021.130196
9. Alrabie H, Alsamadany H, El-Zohri M (2024) Green Synthesis of Phytogenic nanoparticles using Plant Extracts and their Promising Roles in Sustainable Agriculture: A Review. Biosci Res 21(4):738–749.
10. Singh J, Dutta T, Kim KH et al (2018) 'Green'synthesis of metals and their oxide nanoparticles: applications for environmental remediation. J Nanobiotechnol 16(1):84. https://doi.org/10.1186/s12951-018-0408-4
11. Guzo S, Lulekal E, Nemomissa S (2023) Ethnobotanical study of underutilized wild edible plants and threats to their long-term existence in Midakegn District, West Shewa Zone, Central Ethiopia. J Ethnobiol Ethnomed 19(1):30. https://doi.org/10.1186/s13002-023-00601-8
12. Ghosh S, Sarkar T, Chakraborty R (2023) Underutilized plant sources: a hidden treasure of natural colors. Food Biosci 52:102361. https://doi.org/10.1016/j.fbio.2023.102361
13. Knez M, Ranić M, Gurinović M (2024) Underutilized plants increase biodiversity, improve food and nutrition security, reduce malnutrition, and enhance human health and well-being. Let's put them back on the plate! Nutr Rev 82(8):1111–1124. https://doi.org/10.1093/nutrit/nuad103
14. Gupta R, Nigam A, Kapila R (2022) Cultivation and conservation of underutilized medicinal and agricultural plants in India. Proc Natl Acad Sci India B 92(4):741–745. https://doi.org/10.1007/s40011-022-01405-8
15. Banerjee S, Kar P, Sarkar I et al (2022) Structural elucidation and chemical characterization of underutilized fruit silverberry (*Elaeagnus pyriformis*) silver nanoparticles playing a dual role as anti-cancer agent by promoting apoptosis and inhibiting ABC transporters. S Afr J Bot 145:243–257. https://doi.org/10.1016/j.sajb.2021.06.029

16. Eker F, Duman H, Akdaşçi E et al (2024) A comprehensive review of nanoparticles: from classification to application and toxicity. Molecules 29(15):3482. https://doi.org/10.3390/molecules29153482
17. Alshammari BH, Lashin MM, Mahmood MA et al (2023) Organic and inorganic nanomaterials: fabrication, properties and applications. RSC Adv 13(20):13735–13785. https://doi.org/10.1039/D3RA01421E
18. Khan Y, Sadia H, Ali Shah SZ et al (2022) Classification, synthetic, and characterization approaches to nanoparticles, and their applications in various fields of nanotechnology: a review. Catalysts 12(11):1386. https://doi.org/10.3390/catal12111386
19. Altammar KA (2023) A review on nanoparticles: characteristics, synthesis, applications, and challenges. Front Microbiol 14:1155622. https://doi.org/10.3389/fmicb.2023.1155622
20. Joudeh N, Linke D (2022) Nanoparticle classification, physicochemical properties, characterization, and applications: a comprehensive review for biologists. J Nanobiotechnol 20(1):262. https://doi.org/10.1186/s12951-022-01477-8
21. Szczyglewska P, Feliczak-Guzik A, Nowak I (2023) Nanotechnology–general aspects: a chemical reduction approach to the synthesis of nanoparticles. Molecules 28(13):4932. https://doi.org/10.3390/molecules28134932
22. Roy A, Bulut O, Some S et al (2019) Green synthesis of silver nanoparticles: biomolecule-nanoparticle organizations targeting antimicrobial activity. RSC Adv 9(5):2673–2702. https://doi.org/10.1039/C8RA08982E
23. McClements J, McClements DJ (2016) Standardization of nanoparticle characterization: methods for testing properties, stability, and functionality of edible nanoparticles. Crit Rev Food Sci Nutr 56(8):1334–1362. https://doi.org/10.1080/10408398.2014.970267
24. Kaur R, Avti P, Kumar V et al (2021) Effect of various synthesis parameters on the stability of size controlled green synthesis of silver nanoparticles. Nano Express 2(2):020005. https://doi.org/10.1088/2632-959X/abf42a
25. Modena MM, Rühle B, Burg TP et al (2019) Nanoparticle characterization: what to measure? Adv Mater 31(32):1901556. https://doi.org/10.1002/adma.201901556
26. Khan I, Saeed K, Khan I (2019) Nanoparticles: properties, applications and toxicities. Arab J Chem 12(7):908–931. https://doi.org/10.1016/j.arabjc.2017.05.011
27. Kumar A, Dixit CK (2017) Methods for characterization of nanoparticles. In: Advances in nanomedicine for the delivery of therapeutic nucleic acids. Woodhead Publishing, pp 43–58. https://doi.org/10.1016/B978-0-08-100557-6.00003-1
28. Raudone L, Savickiene N (2024) Phytochemical profiles of plant materials: from extracts to added-value ingredients. Plants 13(7):964. https://doi.org/10.3390/plants13070964
29. Perumalsamy H, Balusamy SR, Sukweenadhi J et al (2024) A comprehensive review on *Moringa oleifera* nanoparticles: importance of polyphenols in nanoparticle synthesis, nanoparticle efficacy and their applications. J Nanobiotechnology 22(1):71. https://doi.org/10.1186/s12951-024-02332-8
30. Mohammed GM, Hawar SN (2022) Green biosynthesis of silver nanoparticles from *Moringa oleifera* leaves and its antimicrobial and cytotoxicity activities. Int J Biomater 2022(1):4136641. https://doi.org/10.1155/2022/4136641
31. Jadhav V, Bhagare A, Ali IH et al (2022) Role of *Moringa oleifera* on green synthesis of metal/metal oxide nanomaterials. J Nanomater 2022(1):2147393. https://doi.org/10.1155/2022/2147393
32. Chakraborty A, Haque SM, Dey D et al (2022) Phytogenic silver nanoparticles from tissue-cultured *Kaempferia angustifolia*—an underutilized medicinal herb: a comparative antibacterial study on urinary pathogens. J Genet Eng Biotechnol 20(1):131. https://doi.org/10.1186/s43141-022-00414-4
33. Babu N, Singh A, Singh R (2017) A review on therapeutic potential and phytochemistry of Tephrosia purpurea. Bull Pure Appl Sci-Botany 36(2):91–104
34. Nigam S (2021) *Tephrosia purpurea*: from traditional use to scientific assessment: a review. Environ Sci Ecol: Curr Res 2:1020

35. Shashidhara VM, Srinivasappa PM, H. Jadhav A et al (2023) Phyto-synthesis of silver nanoparticles from *Tephrosia purpurea* and its in-vitro biogenic actions. Appl Organomet Chem 37(12):e7266. https://doi.org/10.1002/aoc.7266
36. Shakoor A, Ferdous UT, Khan SA et al (2025) Green synthesis of gold nanoparticles using *Clerodendrum trichotomum* Thunberg for antibacterial and anticancer applications. Int J Nanomed:2645–2658. https://doi.org/10.2147/IJN.S503254
37. Khan SA, Shahid S, Lee CS (2020) Green synthesis of gold and silver nanoparticles using leaf extract of *Clerodendrum inerme*; characterization, antimicrobial, and antioxidant activities. Biomolecules 10(6):835. https://doi.org/10.3390/biom10060835
38. Ali-Seyed M, Ayesha S (2020) Calotropis-a multi-potential plant to humankind: special focus on its wound healing efficacy. Biocatal Agric Biotechnol 28:101725. https://doi.org/10.1016/j.bcab.2020.101725
39. Kamala NS, Vijayaraghavan K (2020) *Calotropis gigantea* flower extract induced biomimetic synthesis of silver and gold nanoparticles and its application in textile. JNST 6(2):891–893. https://doi.org/10.30799/jnst.300.20060203
40. Kalu AO, Egwim EC, Jigam AA et al (2022) Green synthesis of magnetite nanoparticles using calotropis procera leaf extract and evaluation of its antimicrobial activity. Nano Express 3(4):045004. https://doi.org/10.1088/2632-959X/aca925
41. Periakaruppan R, Romanovski V, Vijai Selvaraj KS et al (2024) Calotropis Gigantea latex-derived zinc oxide nanoparticles: biosynthesis, characterization, and biofunctional applications. Eng 5(3):1399–1406. https://doi.org/10.3390/eng5030073
42. Shanwaz MM, Shyam P (2022) Synthesis of silver nanoparticles from *Vitex negundo* plant by green method and their bactericidal effects. Lett Appl NanoBiosci 12(2):59–62. https://doi.org/10.33263/LIANBS122.059
43. Yayintas OT, Demir N, Canbolat F et al (2024) Characterization, biological activity, and anticancer effect of green-synthesized gold nanoparticles using *Nasturtium officinale* L. BMC Complement Med Ther 24(1):346. https://doi.org/10.1186/s12906-024-04635-7
44. Ilavenil KK, Senthilkumar V, Kasthuri A (2025) Green synthesis of metal nanoparticles from three medicinal plants: a review of environmental and health applications. Discover Catal 2(1):3. https://doi.org/10.1007/s44344-025-00007-6
45. Singh A, Gautam PK, Verma A et al (2020) Green synthesis of metallic nanoparticles as effective alternatives to treat antibiotics resistant bacterial infections: a review. Biotechnol Rep (Amst) 25:e00427. https://doi.org/10.1016/j.btre.2020.e00427
46. Nguyen NT, Nguyen LM, Nguyen TT et al (2022) Formation, antimicrobial activity, and biomedical performance of plant-based nanoparticles: a review. Environ Chem Lett 20(4):2531–2571. https://doi.org/10.1007/s10311-022-01425-w
47. Bukhari A, Ijaz I, Gilani E et al (2021) Green synthesis of metal and metal oxide nanoparticles using different plants' parts for antimicrobial activity and anticancer activity: a review article. Coatings 11(11):1374. https://doi.org/10.3390/coatings11111374
48. Karnwal A, Jassim AY, Mohammed AA et al (2024) Nanotechnology for healthcare: plant-derived nanoparticles in disease treatment and regenerative medicine. Pharmaceuticals 17(12):1711. https://doi.org/10.3390/ph17121711
49. Barathi S, Ramalingam S, Krishnasamy G et al (2024) Exploring the biomedical frontiers of plant-derived nanoparticles: synthesis and biological reactions. Pharmaceutics 16(7):923. https://doi.org/10.3390/pharmaceutics16070923
50. Flieger J, Flieger W, Baj J et al (2021) Antioxidants: classification, natural sources, activity/capacity measurements, and usefulness for the synthesis of nanoparticles. Materials 14(15):4135. https://doi.org/10.3390/ma14154135
51. Ulaş F, Yüksel E, Dinçer D et al (2025) Recent advances in plant-based green synthesis of nanoparticles: a sustainable approach for combating plant-parasitic nematodes. Sustainability 17(9):4152. https://doi.org/10.3390/su17094152

52. Abegunde SM, Afolayan BO, Ilesanmi TM (2024) Ensuring sustainable plant-assisted nanoparticles synthesis through process standardization and reproducibility: challenges and future directions–a review. Sustainable Chem One World 3:100014. https://doi.org/10.1016/j.scowo.2024.100014
53. Aguilar-Pérez KM, Ruiz-Pulido G, Medina DI et al (2022) Regulations and risk assessment of microbial green nanotechnology. In: Handbook of microbial nanotechnology. Academic Press, pp 191–208. https://doi.org/10.1016/B978-0-12-823426-6.00020-6

Chapter 2
Important Bioactive Phytocompounds of Underutilized Plants and Their Derived Nanoparticles

Somvir Singh, Frank Zakeyo Mwandira, Arti Thakur, Priya Sharma, Mitanshi Sharma, Vishav Kiran, Rahul Kumar, Avnish Chauhan, and Muskan

Abstract Underutilized plants which act as a diverse bioactive phytocompounds include flavonoids, alkaloids, terpenoids, and phenolics, which shows meaningful industrial and pharmacological potential. These plants despite their significance in ethnobotany are rich in phytochemical profiles. Recent advancements in green nanotechnology are the nanoparticles that are biosynthesized by the plant extract, result in ecofriendly routes to increase the stability, effectiveness, and bioavailability of these compounds. The nanoparticles derived from the plant extract show various applications such as antimicrobial, antioxidant, anticancer, and anti-inflammatory activities. In this review, we emphasized the importance of phytocompounds and the underutilized species, especially for the synthesis of nanoparticles and for the unlocking of the bush in environmental cleanup, sustainable agriculture, and in nanomedicines. Systematic screening, molecular characterization, and

S. Singh (✉) · R. Kumar · Muskan
Department of Biosciences, University Institute of Biotechnology, Chandigarh University, Gharuan, Mohali, Punjab, India

F. Z. Mwandira
Department of Agriculture Extension, University of Agriculture and Natural Resources, Lilongwe, Malawi

A. Thakur · P. Sharma
Department of Botany, Shoolini Institute of Life Sciences and Business Management, Solan, Himachal Pradesh, India

M. Sharma
Department of Forensic Science, University Institute of Allied Health Sciences, Chandigarh University, Gharuan, Mohali, Punjab, India

V. Kiran
LR Institute of Pharmacy Jabil Kyar, Solan, Himachal Pradesh, India

A. Chauhan
Department of Environmental Science, Graphic Era Hill University, Dehradun, Uttarakhand, India

© The Author(s), under exclusive license to Springer Nature Switzerland AG 2026
E. Nepovimova et al. (eds.), *Underutilized Plants Based-Nanoparticles: Herbal Remedy to Treat Human Diseases*, https://doi.org/10.1007/978-3-032-20384-7_2

biocompatibility on these assessments should be focused in the future research in order to use fully medicinal and technological potential of the botanical resources.

Keywords Underutilized plants · Bioactive phytocompounds · Green synthesis · Plant-based nanoparticles · Nanomedicine

2.1 Introduction

Underutilized crop plants are usually considered as those plant species that are either less accessible, less consumed, or used in the rural area [1]. Many of these as-yet-unknown species have been described as having enormous value, yet have been ignored by plant breeders, scientists, and government decision-makers. Different names, like orphan crops, neglected and underutilized crops, traditional and minor crops, local crops, food of the poor, have been used in the literature, showing the lack of a common definition [2]. Underutilized crops are usually grouped based on end uses, which include beverages, cereals, nuts, vegetables and fruits, and medicinal [3].

2.2 Importance in Food, Pharmaceutical, and Industrial Sectors

Of the estimated 40,000–100,000 plant species that have historically been used for food, fibre, housing, and for other commercial, cultural, and medical purposes, only a small proportion are the major sources of food for the growing world population [4]. The remaining plant diversity is considered as underutilized, despite the fact that global food security and the economic prosperity of rural peoples still rely heavily upon a small number of species. Medicinal and non-medicinal plants have been used for various purposes such as food, fodder, fibre, fuel, nutrition, and for functional foods, natural colours, and flavouring agents and have been published separately. Most of these underutilized wild edible plants have also been reported as minor, neglected, local, orphan, or promising species which are being used as food, oil, fibre, and even some of them are used as medicine since ancient time. Promise of edible plants as a protective food. These crops have not been used to their potential because of limitations such as limited availability, short shelf life, lack of recognition of nutritional potential, low consumer awareness, and stigma related to them as 'famine foods' or 'poor people's food [5]. According to Taylor (2015), pharmaceuticals are biologically active and therapeutic compounds used to cure illnesses. These structures can be simple aromatic compounds or complicated ones that make up medications [6]. Pharmaceuticals come in liquid, gel, pill, or capsule form, and their effectiveness will be reduced if the dosage is not taken within the recommended range. In order to ensure social, environmental, and economic sustainability for future generations, sustainable food systems ensure that everyone has access to food security and nutrition. A number of global phenomena pose a danger to the sustainability of agricultural and food systems, chief among them being the anticipated rise in the world's population. It is predicted that the world's population will increase by more than 30% by 2050, from the present 7.7 billion to over 10 billion [7]. Due to rising greenhouse gas emissions, deforestation, and water usage, this will undoubtedly result in an expansion of agriculture and food production, which will exacerbate ecological inadequacies and climate change [8].

2.3 Significance of Bioactive Phytocompounds

The most significant classes of phytoconstituents found in many plants are flavonoids and phenolic acids, which are also abundant in human diet. Flavonoids are active phytochemicals that contain several phenolic units and are classified as polyphenols. Flavonoids and phenolic acids have demonstrated a number of distinct biological activities that have been linked to the protection of oxidative stress-related illnesses, including cancer, heart disease, and neurological disorders [9].

2.4 Bioactive Phytocompounds in Underutilized Plants

The greatest family of natural products is called terpenoids, or isoprenoids (derived etymologically from turpentine). Hemiterpenoids, monoterpenoids, iridoids, sesquiterpenoids, diterpenoids, sesterterpenoids, triterpenoids, tetraterpenoids, and polyterpenoids are classifications of terpenoids that are typically based on isoprene units (2-methylbuta-1,3-diene; 5 carbon atoms) [10]. Alkaloids are chemical molecules with a nitrogen atom that have biological activity.

These naturally occurring substances are very diverse due to their numerous structural configurations. The several broad groupings of alkaloids include pyrrolidine, pyridine, quinoline, isoquinoline, indole, quinazoline, and so on [11]. Glucosinolates are a class of thioglucosides that are present in many plants, particularly those in the Brassicaceae family, which includes cauliflower, broccoli, cabbage, and brussels sprouts. These are a class of chemical compounds made from glucose and an amino acid that contain nitrogen and sulfur. The enzyme thioglucoside glucohydrolase (myrosinase) is also present in plants that contain glucosinolates. The enzyme myrosinase instantly hydrolyses the parent glucosinolate when it comes into contact with glucosinolates in the presence of water [12]. Numerous research teams have thoroughly examined saponins over the past 10 years in an effort to determine the most effective extraction techniques as well as to clarify their structure, distribution, biosynthesis, commercial significance, and pharmacological significance. Underutilized plants contained various bioactive

Table 2.1 Bioactive compounds in underutilized plants

Sr. No.	Underutilized plants	Bioactive compounds	References
1.	*Amaranthus* spp.	Flavonoids, phenolic acid, tocopherols, squalene	[14]
2.	*Bauhinia variegata*	Kaempferol, quercetin, tannins, saponins	[15]
3.	*Chenopodium quinoa*	Quercetin, kaempferol, saponins	[16]
4.	*Coccinia grandis*	Cucurbitacins, carotenoids, flavonoids	[17]
5.	*Colocasia esculenta*	Phenolics, flavonoids, tannins	[18]
6.	*Eleusine coracana*	Ferulic acid, catechins, proanthocyanidins	[19]
7.	*Fagopyrum esculentum*	Rutin, quercetin, catechins	[20]
8.	*Ipomoea aquatica*	β-carotene, phenolic acids, flavonoids	[21]
9.	*Moringa oleifera*	Quercetin, isothiocyanates, chlorogenic acid	[22]
10.	*Portulaca oleracea*	Omega-3 fatty acids, betalains, flavonoids	[23]
11.	*Tribulus terrestris*	Protodioscin, steroidal saponins	[24]
12.	*Sorghum bicolor*	3-deoxyanthocyanidins, phenolic acids, tannins	[25]

compounds such as flavonoids, saponin, carotenoids, and tocopherol shown in Table 2.1. One of the most valued and promising bioactive groups is saponin, which has a variety of beneficial biological effects. As a result, saponin is currently used in healthcare supplements rather than as a processing aid in commercial applications. In order to have a comprehensive and methodical comprehension of saponin compounds [13].

2.5 Nanotechnology and Its Integration with Phytocompounds

A revolutionary method for enhancing the pharmacological activity of phytochemicals is provided by nanotechnology. Scientists can greatly increase these compounds' solubility, shield them from deterioration, and prolong their circulation time by encasing them in nanoscale carriers (usually 1–100 nm) [26]. Therefore, plant-mediated synthesis avoids the use of toxic chemical reducing and stabilizing agents because plant extracts contain diverse phytochemicals (such as phenolics, flavonoids, terpenoids, alkaloids, and proteins) that naturally act as reducing, capping, and stabilizing agents during nanoparticle synthesis [27]. Utilizing these neglected species with their bioactive compounds not only supports nanotechnology innovation but also promotes biodiversity conservation, new medical treatment, and the valorization of indigenous resources. Plant-mediated nanoparticle synthesis using an underutilized plant with its bioactive compounds responsible for nanoparticle formation is shown in Fig. 2.1. In cancer treatment, plant-derived nanoparticles (PDNPs) provide a number of important benefits, especially in targeted medication administration, gene therapy, and photothermal therapy [28–30].

Fig. 2.1 Schematic representation of plant-mediated nanoparticle synthesis using underutilized plant resources. This figure illustrates the green synthesis pathway for nanoparticles using plant extracts. After preparing the plant material through drying and grinding, extraction is carried out to isolate phytochemicals responsible for nanoparticle formation

2.6 Major Underutilized Plants and Their Key Phytocompounds

2.6.1 Moringa oleifera

M. oleifera Lam., commonly referred to as the miracle tree, horseradish tree, or drumstick tree, belonging to the Moringaceae family, is capable of growing short and slender. It originates from Eastern regions, including the Himalayas, India, Pakistan, Asia Minor, and Africa. Originally native to Arabia, the plant is now found in other countries as well, owing to its resilience. The cultivation in various habitats is possible. The roots have the ability to grow deeply into the soil and retain water for extended durations, enabling the plant to grow in arid and desert soils under varying levels of rainfall. The ideal temperature for *M. oleifera* to thrive ranges from 25 to 30 °C [31].

2.6.1.1 Effect of Antioxidants

Antioxidants are substances that have the capacity to scavenge free radicals, which are the cause of numerous chronic illnesses in humans. The majority of phenolic components, including flavonoids, tannins, and alkaloids, are in charge of antioxidant properties. Studies have examined the antioxidant qualities of both mature and immature *Moringa* leaves [31].

2.6.1.2 Antimicrobial Activity

M. oleifera is being investigated to determine its antibacterial components, just like other medicinal plants. Methanolic leaf extract of *M. oleifera* might inhibit the growth of *Salmonella* typhi (15 mm), *Staphylococcus aureus* (10 mm), *Escherichia coli* (16 mm), and *Bacillus subtilis* (14 mm) [32]. Other factors that contribute to its antibacterial and fungicidal properties include its high flavonoid and polyphenol component content, as well as the presence of chemicals like benzyl isothiocyanate and benzyl glucosinolate [33].

2.6.2 Amaranthus spinosus

Amaranthus, a tropical food plant that dates back to the third millennium, is currently of significant interest. There are about 70 species in the genus *Amaranthus* (L.), which is divided into grain and vegetable amaranths. Only 15 species of amaranth are endemic to Asia, Africa, Australia, and Europe, whereas the majority are native to America [34]. The plant is said to be resistant to heat, drought, and pests and is suited to grow in a variety of agro climatic situations. *Amaranthus* species are found worldwide in temperate, tropical, and subtropical climate zones. The crop proved that it could be grown in regions and during seasons where other crops could not be grown [35].

2.6.2.1 Effect of Antioxidants

Because of their protective functions against reactive oxygen species (ROS), which are created during metabolic reactions in all living things, antioxidants are crucial components of food and medicinal goods. Chronic diseases caused by oxidative stress have been linked to ROS. The antioxidant activity of the aqueous extract of *Amaranthus* spp. derived from various plant sections has been documented in a number of recent antioxidant investigations. The chemical assay-based antioxidant activity of amaranth leaves, seeds, flowers, sprouts, and stalks from several species. The leaves of *A. hypochondriacus* showed the highest antioxidant potential, with ferric reducing antioxidant power (FRAP) and oxygen radical scavenging activity (ORAC) values of 62.2 μmol ascorbic acid equivalent (AAE)/g dry weight (DW) and 451.4 μmol Trolox equivalent (TE)/g DW, respectively [36].

2.6.3 Bauhinia variegata

B. variegata is a flowering plant that is a member of the *Leguminosae* family (subfamily: Caesalpiniaceae). Its local name is "kachnar," and it is also referred to as the orchid tree or mountain ebony. Because of the exquisite beauty of its blossoms, it is grown as a decorative tree. The leftovers of kachnar are roasted and consumed as a delectable dish between March and April. Salivation, sore throat, cough, haemorrhoids, haematuria, menorrhagia, enlargement of the neck glands, and malignancies are all treated with the bark. The dried buds can be used to cure diarrhoea, worms, haemorrhoids, and dysentery, while the bark decoction can help with ulcers and skin conditions [37]. Aqueous extract of leaves is utilized to treat inflammatory diseases and T1 and T2 diabetes [38].

2.6.3.1 Antibacterial Properties

Due to the presence of secondary metabolites such as terpenoids, phenolics, flavonoids, anthraquinones, saponins, tannins, and alkaloids in *B. variegata*, ethyl alcohol leaf extract of *B. variegata* showed antibacterial activity against various microorganisms viz. *K. pneumoniae*, *Proteus* spp., *E. coli*, and *Pseudomonas* spp. showed zone of inhibition 11.33 mm, 10.33 mm, 11.67 mm, and 11.67 mm, respectively [39].

2.7 Green Synthesis Approach

Green synthesis of nanoparticles has emerged as an alternative to conventional chemical and physical approaches because it is simple, cost-effective, and environmentally friendly. Plant-mediated synthesis, in particular, avoids toxic reducing and stabilizing agents by exploiting the diverse phytocompounds found in plant extracts [40]. Utilizing these neglected species not only supports nanotechnology innovation but also promotes biodiversity conservation and the valorization of indigenous resources.

2.7.1 Types of Nanoparticle Synthesis

NPs have gained attention due to their unique physicochemical, optical, electronic, and catalytic properties that differ significantly from their bulk counterparts. The methods are physical, chemical, and biological (green), which are broadly used to create nanoparticles. Gold (AuNPs), silver (AgNPs), and iron oxide (Fe_3O_4-NPs) are mostly studied among all nanoparticles that are produced using different methods.

2.7.1.1 Silver Nanoparticles (AgNPs)

The well-known antifungal, antimicrobial, and catalytic applications are silver (AgNPs).

Chemical reduction: It is the most common technique, where the reduction of silver salts ($AgNO_3$) is achieved by using a different chemical agent, such as citrate and sodium borohydride. Surfactants are frequently added to this technique to stabilize the generated nanoparticles [41].

Green synthesis: Phytochemicals (tannins, alkaloids, flavonoids), which act as stabilizing and reducing agents, are provided by plant extracts, microorganisms, and biopolymers [42].

Bioactive Compounds Involved in AgNPs

Flavonoids
Flavonoids, called polyphenolic substances, are catechin, kaempferol, and quercetin. It donates electrons to Ag+ ions which leads to the formation of AgNPs and their reduction. To keep nanoparticles stable, the hydroxy groups serve as binding sites [43].

Phenolic Acids and Tannins
The natural antioxidants and reducing agents' phenolic compounds (caffeic acid, gallic acid, and tannic acid) promote electron transfer, which helps AgNPs nucleate and grow. By making protective layers around the nanoparticles, tannins in particular stop agglomeration [44].

Terpenoids and Alkaloids
Limonene and eugenol are terpenoids which are liable for reducing Ag^+ ions by oxidizing their aldehyde and hydroxyl functional groups. Nitrogen atoms in alkaloids discovered in plants like *Lantana camara* and *Calotropis gigantea* coordinate with metal ions, to stabilize nanoparticles [45].

Examples of Underutilized Plants Used
For the reduction of silver nitrate into AgNPs of ~20–50 nm, the leaf extract of *L. camara* was used. In Gram-negative bacteria (*E. coli*), the nanoparticles show strong antibacterial activity, which proposes their use in biomedicine [45–47]. Plasmon resonance characteristics have a special surface that helps in imaging, biosensing, and drug delivery.

2.7.1.2 Gold Nanoparticles (AuNPs)

The metabolites that donate electrons to decrease Au^{3+} to Au_4 include phenolic acids and flavonoids (such as luteolin and quercetin). Alkaloids and terpenoids use the functional groups like -OH, -C=O, and $-NH_2$ to stabilize AuNPs. Protein and polysaccharides also form a shielding layer that stops nanoparticles from clumping together [40]. AuNPs, antioxidants, and anticancer values are increased by phytochemicals that attach to their surface.

Examples of Underutilized Plants Used

Azadirachta indica (Neem): It contains nimbolide, azadirachtin, and flavonoids which reduce the Au^{3+} to Au^0 and produce AuNPs (~20–50 nm) which are spherical and have cytotoxic and antibacterial activities [48].

2.7.1.3 Zinc Oxide Nanoparticles (ZnONPs)

ZnO nanoparticles exhibits photocatalytic, antibacterial, and UV-blocking properties. By using precipitation method in which the zinc salts (e.g. zinc acetate) in alkaline medium are precipitated and then calcified to form ZnONPs [49]. While in sol–gel method the exact control over the size and shape of particles is by using condensation and hydrolysis of zinc alkoxides. Green synthesis: ZnO nanoparticles are formed into Zn^{2+} at low toxicity where the extract of plants supplies phytochemicals for stabilization and reduction [50].

Role of Bioactive Compounds

Extract of plants that are rich in antioxidants and organic acids is used to create ZnONPs, which are semiconductor nanomaterials. Flavonoids (catechin, quercetin) act as a reducing agent which converts zinc nitrate ($Zn(NO_3)_2$) into zinc hydroxide ($Zn(OH)_2$), which is further calcined to form ZnO. Saponin and terpenoids, by chelating zinc ion, provide surface stability [51].

Examples of Underutilized Plants

Parthenium hysterophorus: ZnONPs showed strong antibacterial activity from the weeds rich in flavonoids, parthenin, and phenolics [51, 52].

Calotropis procera: This contains alkaloids and terpenoids which convert Zn^{2+} to ZnONPs and show UV and antifungal properties [50].

Case Study 2: *Parthenium hysterophorus* for ZnONPs.

From the floral extract of *Parthenium hysterophorus*, ZnONPs are formed, which show antibacterial activity against *Staphylococcus aureus*. This may conclude how weeds may be used for medicinal and environmental purposes [52].

Case Study 3: *Annona squamosa* (custard apple) for ZnONPs.

Leaf extract of *Annona squamosa* synthesized ZnONPs. The photocatalytic destruction of methylene blue dye was shown by resultant nanoparticles which indicates the potential for wastewater treatment applications [51].

2.7.1.4 Iron Oxide Nanoparticles (Fe_3O_4NPs)

Iron oxide nanoparticles are extensively used in magnetic resonance imaging (MRI), drug delivery, and environmental remediation.

Co-precipitation method: Most widely used, where Fe^{2+} and Fe^{3+} salts are precipitated with alkali under inert conditions to produce magnetite nanoparticles [53].

Thermal decomposition: Involves decomposition of organometallic precursors at high temperatures with surfactants, producing highly crystalline NPs [54].

Green synthesis: Plant metabolites and microbial systems can reduce iron salts into magnetite NPs with biocompatible coatings [55].

Role of Bioactive Compounds
Iron oxide nanoparticles are synthesized by reducing ferric (Fe^{3+}) or ferrous (Fe^{2+}) ions using phytochemicals.

Polyphenols, ascorbic acid, and reducing sugars convert Fe^{3+} to Fe^{2+} and subsequently form Fe_3O_4 nanoparticles.

2.8 Characterization Techniques of Nanoparticles

2.8.1 UV–Visible Spectroscopy (UV–Vis)

Surface plasmon resonance (SPR), which is characteristic of nanoparticles in colloidal solutions like silver, gold, and UV-vis spectroscopy, is used to confirm the formation and stability of nanoparticles. The position and width of the absorption bands peak may provide information about the shape, size, stability, and aggregation [56].

2.8.2 X-Ray Diffraction (XRD)

The phase and crystalline structure of nanoparticles are examined by using XRD. Bragg's rule is used to examine the crystal system from the diffraction patterns, and the Debye-Scherer equation is used to calculate the average crystalline size [57]. The flaws and lattice strain are shown by XRD, and the material is amorphous and crystalline.

2.8.3 Scanning Electron Microscopy (SEM) and Transmission Electron Microscopy (TEM)

When it is along energy-dispersive X-ray spectroscopy (EDS), scanning electron microscopy (SEM) may confirm elemental composition and deliver information on the surface morphology, size distribution, and shape of nanoparticles at high magnifications [58].

The crystallinity and interior structures are examined by TEM. Data from X-ray diffraction (XRD) is magnified by rich crystallographic information provided by high-resolution TEM (HRTEM) and selected area electron diffraction (SAED) [59].

2.8.4 Fourier Transform Infrared Spectroscopy (FTIR)

FTIR spectroscopy is used to identify the functional groups and surface chemistry in nanoparticles, confirm the activity of biomolecules, polymers, or surfactants as stabilizing and capping agents throughout the synthesis [60]. Some of the vibrational modes that show bonding contacts on the surface of the nanoparticles and are revealed by spectra [61].

2.9 Biological Activities of Underutilized Plant-Derived Nanoparticles

Plant-derived nanoparticles have gained a lot of attention because of their multifunctional biological properties, biocompatibility, and eco-friendly development. The natural reducing and carping agents such as phenolics, tannins, alkaloids, terpenoids, and flavonoids which help in increasing the biological potential of these nanoparticles [60, 62, 63]. Gold (AuNPs), silver (AgNPs), iron oxide (FeONPs), and zinc oxide (ZnOPs) are obtained from medicinal and underutilized plants among the most studied. Hier biological activities are described in the following sections.

2.9.1 Antioxidant Activity

Phytochemicals restrict on nanoparticles rise of free radical scavenging capacity, which helps in treating the disorders which are linked with oxidative stress. The EC20 value for silver nanoparticle derived from *Camellia sinensis* extract is 25.3 μg/ml reviled that superior antioxidant activity compared to the crude extract value that is 37 μg/ml DPPH assay [40]. This recommends a cooperative relationship with polyphenolic substances. Superoxide and hydroxyl radicals were completely scavenged by zinc oxide nanoparticles (ZnONPs) which are obtained from the leaf extract of *Annona squamosa*, and show complete activity at 50–200 μg/ml [51].

2.9.2 Cytotoxicity and Anticancer Potential

Through a number of important mechanisms, including membrane interactions, apoptosis, and the induction of reactive oxygen species (ROS) production, silver nanoparticles (AgNPs) made with *Calotropis gigantea* demonstrate strong cytotoxic and anticancer properties. Although the enhanced permeability and retention (EPR) effect is a general phenomenon that facilitates their targeted delivery to tumour tissues, the other mechanisms mentioned above are primarily responsible for the specific toxicity at the cellular level [64]. AgNPs from *C. gigantea*, the nanoparticles show the dose-dependent cytotoxicity against HeLa cervical cancer cells in MTT assays, with an IC_{20} value of roughly 0.85 mg/ml with the concentration of 0.2–1.0 mg/ml. Their accuracy was emphasized by the low toxicity of control treatments. *A. indica* (neem) displayed completely anti-cancerous activity against breast cancer (MCF-7) cells, the mitochondrial apoptotic pathways activating. At 50–100 μg/mL, AuNPs induced dose-dependent cytotoxicity, achieving ~50% cell inhibition at ~80 μg/mL [48]. AuNPs confirm the efficiency at non-toxic levels. Positive control (doxorubicin) displayed ~90% confirming inhibition. ZnONPs synthesized using *M. oleifera* leaf extract demonstrated selective cytotoxicity toward HepG2 liver carcinoma cells, showing activity at 25–50 μg/mL while sparing normal cells [65].

2.10 Applications

Beyond biomedical uses, plant-derived nanoparticles offer environmental and industrial benefits.

2.10.1 *Waste-Water Treatment*

ZnONPs and AgNPs synthesized from plant extracts have been used for photocatalytic degradation of industrial dyes such as methylene blue and rhodamine B. At catalytic loadings of 0.1–1 g/L, over 90% dye degradation within 120 minutes under UV exposure has been documented [51]. Underutilized plants used in wastewater treatment include *L. camara* (invasive shrub) which is a source of AgNPs and AuNPs applied through degradation of azo dye and antibacterial wastewater treatment [45]; *Moringa oleifera* seed extract (underutilized protein-rich seed, a source of AgNPs) is applied in bacterial decontamination of wastewater.

2.10.2 *Agriculture*

Nanoparticles from *M. oleifera* act as nanofertilizers and nanopesticides, enhancing nutrient efficiency and suppressing plant pathogens at foliar application rates of 10–50 mg/L [66]. These green nano products support sustainable agro-ecology.

2.10.3 Industrial Formulations

AgNP-coated textiles and packaging materials resist microbial growth, while ZnONPs are utilized in sunscreens for UV protection [62]. Green synthesis pathways reduce environmental toxicity in manufacturing. Examples of underutilized plants in industrial use include *Tamarindus indica* seed source of AuNPs applied in biosensors for detecting pollutants *Punica granatum* peel source of AgNPs applied in food packaging and antimicrobial coatings.

2.11 Conclusion

The present chapter focused on various bioactive compounds present in underutilized plants and their derived nanoparticles. These bioactive compounds responsible for the medical efficacy of plants. Flavonoids, phenolic acids, terpenoids, alkaloids, tannin, carotenoids, and saponin considered as major bioactive constituents in plants. *A. indica* contains azadirachtin, nimbolide, and flavonoids as bioactive compounds that yield spherical AuNPs with antibacterial and cytotoxic activities. *Calotropis procera* exhibited terpenoids and alkaloids that are used to synthesize ZnONPs showing antifungal and UV-protection activity. Plant-derived nanoparticles represent a convergence of nanotechnology and green chemistry, offering multifunctional properties ranging from antimicrobial to anticancer activities, as well as environmental applications. With the use of *M. oleifera*, nanoparticles are produced which work as nanopesticides and nanofertilizers, and inhibit plant pathogens and strengthen nutrient efficiency. AuNPs and AgNPs are synthesized from *L. camara* and used by antimicrobial wastewater treatment and azo dye degradation. They are ideal for developing industrial, biomedical, and environmental automation because of their ecofriendly, cost effectivity, and phytochemical-based stability.

Author's Contribution SS—Writing original draft; FZM, AT, MS, VK—Data interpretation, conceptualization; SS, AT—Final formatting, final approval; RK, AC, PK, MK—Data collection.

Conflict of Interest The authors have no conflict of interest.

Ethical Approval The authors declare that the work presented in this chapter complies with all relevant ethical standards. No human or animal subjects were used in this study.

References

1. Padulosi S, Mal B, King O et al (2015) Minor crops as a central element for sustainable enhanced incomes, empowerment and nutrition in rural India. J Sustain 7:8904–8933. https://doi.org/10.3390/su7078904
2. Mabhaudhi T, O'Reilly P, Walker S (2016) The role of underutilised crops in southern African farming systems: a scope study; Centre for Coordination of Agricultural Research and Development for Southern Africa. J Agric 8:311. https://doi.org/10.3390/su8040302
3. Mabhaudhi T, O'Reilly P, Walker S (2015) The role of underutilised crops in southern African farming systems: a scoping study. J Sustain 5:123–129. https://doi.org/10.3390/su8040302
4. Magbagbeola JA, Adetoso JA, Owolabi OA (2010) Neglected and underutilized species (NUS): panacea for community focused development to poverty alleviation/poverty reduction in Nigeria. J Econ Int Fin 2:208–211. http://www.academicjournals.org/JEIF
5. Williams JT, Haq N (2002) Global research on underutilized crops: an assessment of current activities and proposals for enhanced cooperation. J Bio Div Int 14(6):304–309
6. Taylor D (2015) The pharmaceutical industry and the future of drug development. Issues Environ Sci Tech 41. https://doi.org/10.1039/9781782622345-00001
7. Hazen HD, Alberts HC, Zaniewski KJ (2023) Population geography: social justice for a sustainable world, vol 1. Routledge, p 352. https://doi.org/10.4324/9781003143253
8. Food and Agriculture Organization of the United Nations (2017) The future of food and agriculture: trends and challenges
9. Goleniowski M, Bonfill M, Cusido R et al (2013) Phenolic acids. In: Natural products. Springer, pp 1951–1973. https://doi.org/10.1007/978-3-642-22144-6_64
10. Ludwiczuk A, Skalicka-Wozniak K, Georgiev MI (2017) Terpenoids A2-Badal, Simone. In: Delgoda R (ed) Pharmacognosy. Academic Press, Boston, pp 233–266. https://doi.org/10.1016/B978-0-12-802104-0.00011-1
11. Azimova SS (2013) Natural compounds: plant sources, Structure and Properties. Springer, New York
12. Vig AP, Rampal G, Thind TS (2009) Bio-prospective effects of glucosinolates-a review. LWT-Food Sci Technol 42:1561–1572. https://doi.org/10.1016/j.lwt.2009.05.023
13. Aziz M, Said AO, Melad AS (2019) A review on saponins from medicinal plants: chemistry, isolation, and determination. J Nanomed Res 8(1):6–12
14. Rastogi A, Shukla S, Amarnath (2013) A new millennium crop of nutraceutical values. Crit Rev Food Sci Nutr 53(2):109–125. https://doi.org/10.1080/10408398.2010.517876
15. Singh B et al (2004) Phytochemical screening of Bauhinia variegata. J Ethnopharmacol 92:267–272. https://doi.org/10.1186/s43094-021-00251-1
16. Gordillo-Bastidas E (2016) Nutritional properties of quinoa. Food Sci Technol 36(1):1–7
17. Reddy MK (2005) Antioxidant properties of Coccinia grandis. J Ethnopharmacol 99:63–71. https://doi.org/10.2174/1381612826666200928160357
18. Huang CC (2007) Phenolic compounds of taro. Food Chem 105:528–536. https://doi.org/10.3389/fnut.2022.884844
19. Chetan S, Malleshi NG (2007) Finger millet phenolic compounds. Food Chem 105:862–870. https://doi.org/10.5923/j.env.20160602.03
20. Kreft I (2006) Buckwheat antioxidants. Plant Foods Hum Nutr 61:1–4. https://doi.org/10.1017/S0954422415000190
21. Islam S (2007) Nutritional properties of *Ipomoea aquatica*. Int J Food Sci Nutr 58:319–330. https://doi.org/10.3329/aajbb.v2i2.64384
22. Leone A (2015) Nutritional and nutraceutical profile of *Moringa oleifera* leaves. Int J Mol Sci 16(12):12791–12835. https://doi.org/10.3390/ijms160818923
23. Uddin MK (2014) Purslane: nutrient and bioactive profile. Plant Food Hum Nutr 69:104–111
24. Adaikan PG (2001) Phytochemical and pharmacological profile of Tribulus terrestris. Phytomedicine 8:103–108
25. Awika JM, Rooney LW (2004) Sorghum phytochemicals. J Agric Food Chem 52:2348–2358

26. Parvin N, Kumar V, Joo SW et al (2022) Recent advances in the characterized identification of mono-to-multi-layer graphene and its biomedical applications: a review. Electronics 11:3345. https://doi.org/10.3390/electronics11203345
27. Ahmed S, Saifullah Ahmad M, Swami BL et al (2016) Green synthesis of silver nanoparticles using Azadirachta indica aqueous leaf extract. J Radiat Res Appl Sci 9(1):1–7. https://doi.org/10.1016/j.jrras.2015.06.006
28. Ahmar S, Mahmood T, Fiaz S et al (2021) Advantage of nanotechnology-based genome editing system and its application in crop improvement. Front Plant Sci 12:663849. https://doi.org/10.3389/fpls.2021.663849
29. Zhao L, Lu L, Wang A et al (2020) Nano-biotechnology in agriculture: use of nanomaterials to promote plant growth and stress tolerance. J Agric Food Chem 68(7):1935–1947. https://doi.org/10.1021/acs.jafc.9b06615
30. Chowdhury MA, Iqbal MZ, Rana MM et al (2024) Green synthesis of novel green ceramic-based nanoparticles prepared by sol-gel technique for diverse industrial application. Results Surf Interfaces 14:100178. https://doi.org/10.1016/j.rsurfi.2023.100178
31. Chaudhary K, Chaurasia S (2017) Neutraceutical properties of *Moringa oleifera*: a review. Eur J Pharm Med Res 4:646–655
32. El-Sherbiny GM, Alluqmani AJ, Elsehemy IA et al (2024) Antibacterial, antioxidant, cytotoxicity, and phytochemical screening of *Moringa oleifera* leaves. Sci Rep 14(1):30485. https://doi.org/10.1038/s41598-024-80700-y
33. Bako SS, Okere JU, Etonihu AC et al (2010) Medicinal values of Moringa oleifera leave extracts. In: Raw materials research and development council: Moringa – A national crop for economic growth and development, pp 107–114
34. Peter K, Gandhi P (2017) Rediscovering the therapeutic potential of Amaranthus species: a review. Egypt J Basic Appl Sci 4:196–205. https://doi.org/10.1016/j.ejbas.2017.05.00
35. Martinez-Lopez A, Millan-Linares MC, Rodriguez-Martin NM et al (2020) Nutraceutical value of kiwicha (*Amaranthus caudatus* L.). J Funct Foods 65:103735. https://doi.org/10.1016/j.jff.2019.103735
36. Li H, Deng Z, Liu R et al (2015) Characterization of phenolics, betacyanins and antioxidant activities of the seed, leaf, sprout, flower and stalk extracts of three Amaranthus species. J Food Compos Anal 37:75–81. https://doi.org/10.1016/j.jfca.2014.09.003
37. Kumar S, Baniwal P, Kaur J et al (2020) Kachnar (*Bauhinia variegata*). In: Antioxidants in fruits: properties and health benefits, vol 16. Springer Singapore, Singapore, pp 365–377. https://doi.org/10.63793/ISRE/0010
38. Nariyal V, Sharma PK (2017) Kanchnar (Bauhinia Variegata) as a medicinal herb: a systematic review. Int J Adv Res (Indore) 5(9):587–591. https://doi.org/10.26538/tjnpr/v2i1.2
39. Mishra A, Sharma AK, Kumar S et al (2013) *Bauhinia variegata* leaf extracts exhibit considerable antibacterial, antioxidant, and anticancer activities. Biomed Res Int 2013(1):915436. https://doi.org/10.1155/2013/915436
40. Ahmed S, Ahmad M, Swami BL et al (2016) A review on plants extract mediated synthesis of silver nanoparticles for antimicrobial applications: a green expertise. J Adv Res 7(1):17–28. https://doi.org/10.1016/j.jare.2015.02.007
41. Khan I, Saeed K, Khan I (2019) Nanoparticles: properties, applications and toxicities. Arab J Chem 12(7):908–931. https://doi.org/10.1016/j.arabjc.2017.05.011
42. Ibrahim HM (2015) Green synthesis and characterization of silver nanoparticles using banana peel extract and their antimicrobial activity against representative microorganisms. J Rad Res Appl Sci 8(3):265–275. https://doi.org/10.1016/j.jrras.2015.01.007
43. Ahmed S, Saifullah AM, Swami BL et al (2016) Green synthesis of silver nanoparticles using *Azadirachta indica* aqueous leaf extract. J of Rad Res and Appl Sci 9(1):1–7. https://doi.org/10.1016/j.jrras.2015.06.006
44. Iravani S (2011) Synthesis of silver metal nanoparticles using plants. Green Chem 11(13):2608–2678. https://doi.org/10.1039/C1GC15386B

45. Balashanmugam P, Kalaichelvan PT (2015) Biosynthesis characterization of silver nanoparticles using *Lantana camara* leaf extract and their antimicrobial activity. App Nano 5(3):305–311. https://doi.org/10.1007/s13204-014-0314-7
46. Turkevich J, Stevenson PC, Hillier J (1951) A study of the nucleation and growth processes in the synthesis of colloidal gold. Dis Faraday Soc 11:55–75. https://doi.org/10.1039/DF9511100055
47. Nikoobakht B, El-Sayed MA (2003) Preparation and growth mechanism of gold nanorods using seed-mediated growth method. Chem Mater 15(10):1957. https://doi.org/10.1021/cm020732l
48. Sharma VK, Yngard RA, Lin Y (2012) Silver nanoparticles: green synthesis and their antimicrobial activities. Adv Colloid Interface Sci 145(1–2):83–96. https://doi.org/10.1016/j.cis.2008.09.002
49. Kavitha C, Rajeshkumar S, Malarkodi C (2021) Green synthesis of zinc oxide nanoparticles using plant extracts: a review. Environ Nanotechnol Monit Manag 16:100467. https://doi.org/10.1016/j.enmm.2021.100467
50. Sangeetha G, Rajeshwari S, Venckatesh R (2011) Green synthesis of zinc oxide nanoparticles by Aloe vera extract: structure and optical properties. Mat Res Bull 46(12):2560–2566. https://doi.org/10.1016/j.materresbull.2011.07.046
51. Elumalai K, Velmurugan S, Ravi S et al (2015) Green synthesis of zinc oxide nanoparticles using Annona squamosa leaf extract and its antibacterial activity. Spectrochim Acta A Mol Biomol Spectrosc 149:213–219. https://doi.org/10.1016/j.saa.2015.04.066
52. Sharma J, Kaith BS, Kalia S (2019) Green synthesis of ZnO nanoparticles using Parthenium hysterophorus leaf extract and evaluation of their antibacterial properties. J Mater Sci: Mater Med 30(6):1–10. https://doi.org/10.1007/s10856-019-6281-2
53. Massart R (1981) Preparation of aqueous magnetic liquids in alkaline and acidic media. IEEE Trans Magn 17(2):1247–1248. https://doi.org/10.1109/TMAG.1981.1061188
54. Sun S, Zeng H (2002) Size-controlled synthesis of magnetite nanoparticles. J Am Chem Soc 124(28):8204–8205. https://doi.org/10.1021/ja026501x
55. Akbarzadeh A (2012) Synthesis and characterization of magnetic nanoparticles for biomedical applications. J Nanomedicine 8(12):1–8. https://doi.org/10.2147/NSA.S24328
56. Kumar V, Yadav SK (2009) Plant-mediated synthesis of silver and gold nanoparticles and their applications. J Chem Tech Biotechnol 84(2):151–157. https://doi.org/10.1002/jctb.2023
57. Cullity BD, Stock SR (2014) Elements of X-ray diffraction, 3rd edn. Pearson Education
58. Goldstein JI, Newbury DE, Echlin PJ et al (2003) Special topics in electron beam x-ray microanalysis. In: Scanning electron microscopy and X-ray microanalysis: third edition. Springer US, Boston, MA, pp 453–536
59. Williams DB, Carter CB (2009) Transmission electron microscopy: a textbook for materials science, 2nd edn. Springer
60. Iravani S (2011) Green synthesis of metal nanoparticles using plants. Green Chem 13(10):2638–2650. https://doi.org/10.1039/c1gc15386b
61. Jain D, Mehata MS (2017) Medicinal plant leaf extract and pure flavonoid-mediated green synthesis of silver nanoparticles and their enhanced antibacterial property. Sci Rep 7(1):15867. https://doi.org/10.1038/s41598-017-15724-8
62. Rai M, Yadav A, Gade A (2014) Silver nanoparticles as a new generation of antimicrobials. Biotech Adv 27(1):76–83. https://doi.org/10.1016/j.biotechadv.2008.09.002
63. Lara HH, Ayala-Nunez NV, Ixtepan-Turrent L et al (2010) Mode of antiviral action of silver nanoparticles against HIV-1. J Nanobiotechnol 8(1):1–10. https://doi.org/10.1186/1477-3155-8-1
64. Gurunathan S, Park JH, Han JW et al (2013) Comparative assessment of the apoptotic potential of silver nanoparticles synthesized by *Bacillus tequilensis* and *Calotropis gigantea* plant extract in MDA-MB-231 human breast cancer cells: targeting p53 for anticancer therapy. Int Journal of Nanomed 8:4391–4408. https://doi.org/10.2147/IJN.S51871

65. Premanathan M, Karthikeyan K, Jeyasubramanian K et al (2011) Selective toxicity of ZnO nanoparticles toward Gram-positive bacteria and cancer cells by apoptosis through lipid peroxidation. Nanomed Nanotechnol Biol Med 7(2):184–192. https://doi.org/10.1016/j.nano.2010.10.001
66. Kah M, Kookana RS, Gogos A et al (2018) A critical evaluation of nanopesticides and nanofertilizers against their conventional analogues. Nat Nanotechnol 13(8):677–684. https://doi.org/10.1038/s41565-018-0131-1

Chapter 3
Underutilized Plants Derived Metallic and Bimetallic Nanoparticles Against Human Illness

Arti Thakur, Somvir Singh, Kanika Dulta, Priya Sharma, Mehak Pawar, and Anjli Sharma

Abstract A promising area in nanomedicine is the investigation of underused plants as sustainable resources for ecologically approachable synthesis of metallic and bimetallic nanoparticles. These plants have distinct phytochemical profiles full of alkaloids, flavonoids, terpenoids, and phenolic chemicals that act as organic reducing and stabilizing agents during manufacture. Because of their improved physicochemical characteristics, such as high surface area, catalytic efficiency, and stability, the resultant nanoparticles, especially metallic and bimetallic structures, have shown important biomedical applications. Compared to their monometallic counterparts, bimetallic nanoparticles exhibit higher efficacy in targeting human pathogens and cancer cells due to their synergistic effect, which combines the capabilities of two metals. This chapter examines developments in the environmentally friendly synthesis of metallic and bimetallic nanoparticles from underutilized species and assesses potential uses in biomedicine, modes of action, and human health in the future. The difficulties associated with clinical translation, biological activities, and characterization methods are given attention. Utilizing the underutilized plant-derived nanoparticles provides a novel and sustainable approach to treating a wide range of human illnesses.

Keywords Underutilized plants · Green synthesis · Metallic nanoparticles · Bimetallic nanoparticles · Nanomedicine · Human diseases

---

A. Thakur (✉) · P. Sharma · M. Pawar · A. Sharma
Department of Botany, Shoolini Institute of Life Sciences and Business Management, Solan, Himachal Pradesh, India

S. Singh
Department of Biosciences, University Institute of Biotechnology, Chandigarh University Gharuan, Mohali, Punjab, India

K. Dulta
Center of Advanced Innovation, VSB-Technical University of Ostrava, Ostrava-Poruba, Czech Republic

© The Author(s), under exclusive license to Springer Nature Switzerland AG 2026

E. Nepovimova et al. (eds.), *Underutilized Plants Based-Nanoparticles: Herbal Remedy to Treat Human Diseases*, https://doi.org/10.1007/978-3-032-20384-7_3

3.1 Introduction

Nanotechnology has led to substantial advancements in various scientific and technological fields, particularly in medicine and physiology. In these applications, materials and devices can be engineered to interact with cells and tissues at the molecular (subcellular) level with a high degree of functional specificity. This capability enables a level of integration between technology and biological systems that was previously unattainable [1]. It is essential to recognize that nanotechnology is not a standalone scientific discipline; rather, it represents a convergence of traditional sciences such as chemistry, physics, materials science, and biology, combining the necessary collective expertise to create these innovative technologies. The importance of nanoscience and emerging nanotechnologies for human health in medicine lies in their potential to enhance imaging, sensing, targeted drug delivery, and gene delivery systems [2–5]. By using detrimental byproducts, the creation of toxic solvents and imperfections in surface structure are just a few of the challenges associated with these costly processes. Due to the intricacies and unpredictability of their composition, chemical procedures often involve multiple chemical species or molecules, which can increase toxicity and reactivity, posing risks to both living and non-living organisms. As a result, researchers are increasingly exploring biological methods as a more practical alternative to harmful, expensive, and energy-intensive chemical processes [6]. Thus, an alternative to physico-chemical methods for producing nanoparticles is the green synthesis of nanomaterials, which is favored over those made through physico-chemical processes. On the basis of nanoparticles' (NPs) unique properties, using biological methods is preferred [7].

3.2 Underutilized Plants: A Hidden Resource

Underutilized plants refer to species that are region-specific, rarely cultivated, infrequently consumed, and generally less available. They further pointed out that the neglect of such species by policy makers, researchers, and breeders stems from the fact that their economic and nutritional importance has not yet been fully realized [8]. Several terms such as orphan crops, neglected crops, indigenous crops, traditional, minor, local crops, and poor people's food have been used to describe underutilized species [9]. These species show a vigorous part in local food making and consumption, can thrive in diverse agroecological zones and marginal areas, yet remain largely absent from national policies and mainstream agendas. The high content of carotenoids and minerals in neglected underutilized species enables them to significantly improve dietary micronutrients for millions of people. These species are frequently used by farmers as food supplements during periods of staple crop failure and emergency situations [10, 11] (Table 3.1).

Table 3.1 List of some underutilized plants [12–14]

Plant	Scientific name	Family	Region/country
Jamun	*Syzygium cumini*	Myrtaceae	India, Sri Lanka, Bangladesh
Ber	*Ziziphus mauritiana*	Rhamnaceae	Central Asia and India
Bael	*Aegle marmelos*	Rutaceae	India, Nepal, Bangladesh
Jackfruit	*Artocarpus heterophyllus*	Moraceae	Western Ghats of India
Karonda	*Carrisa carandus*	Apocynaceae	Indo-Malayan region
Carambola	*Averrhoa carambola*	Oxalidaceae	Southeast Asia
Phalsa	*Grewia subinaequalis*	Tiliaceae	Indian subcontinent
Loquat	*Eribotrya japonica*	Rosaceae	South-central China
Tamarind	*Tamarindus indica*	Leguminosae	Tropical Africa
Rambutan	*Nephelium lappaceum*	Sapindaceae	Malay-Indonesian region
Haritaki	*Terminalia chebula*	combretaceae	Indian subcontinent
Green Amaranth	*Amaranthus viridis*	Amaranthaceae	Central and South America
Tassel hyacinth	*Leopoldia comosa*	Hyacinthaceae	Turkey and Mediterranean region
Purslane	*Portulaca oleracea*	Portulacaceae	Cosmopolitan (Mexico, Europe)
Sausage tree	*Kigelia africana*	Bignoniaceae	Tropical Africa
Veldt grape	*Cissus quadrangularis*	Vitaceae	Tropical Asia and Africa
Little millet	*Panicum sumatrense*	Poaceae	South and Southeast Asia

3.3 Phytochemical Richness and Potential for Nanoparticle Synthesis

Underutilized plants are increasingly recognized as valuable reservoirs of phytochemicals due to their adaptation to marginal environments and unique metabolic pathways [15]. Underutilized plants accumulate diverse groups of secondary metabolites for instances flavonoids, phenolic acids, alkaloids, saponins, and terpenoids which contribute to their stress tolerance and potential therapeutic value [16]. For example, phenolic compounds, including chlorogenic, caffeic, and ferulic acids, are widely reported in *Amaranthus viridis* and *A. spinosus* due to which they exhibit strong antioxidant properties [17]. Terpenoids and steroids are characteristic of *Cissus quadrangularis* (an underutilized medicinal plant), which has been used traditionally for bone-healing [18]. Marappan et al. [19] mentioned that underutilized fruit crops, such as minor berries and indigenous tropical fruits, also provide significant amounts of polyphenols, carotenoids, and vitamins, which are critical for nutritional security and functional food development. Phoswa and Mokgalaboni [20] found that *Amaranthus* extract, rich in phenolic acid and betalains, has been used for AgNP (silver nanoparticle) synthesis, producing stable and spherical nanoparticles with antimicrobial activity (Table 3.2).

Table 3.2 Nanoparticle synthesis mediated by plants and their applications

Plant species	Extract used	Nanoparticle type	Reported applications	References
Sambucus nigra L.	Plant extract	Gold nanoparticles (AuNPs)	Lowers blood glucose level, reduces oxidative stress and inflammation	[21]
Moringa oleifera	Plant extract	Iron-zinc oxide nanoparticles	Remarkable electrochemical properties, offering promise for energy utilization	[22]
Camellia sinensis	Plant extract	Palladium nanoparticles (PdNPs)	Catalysis in biaryl synthesis (Suzuki cross-coupling); reduction of 4-nitrophenol	[23]
Coccinia grandis	Stem extract	Silver nanoparticles	Exhibits antipyretic, analgesic, and antidiabetic properties	[24]
Zizyphus mauritiana	Fruit extract	Silver chloride nanoparticles	Possess antibacterial, antifungal, and antiproliferative mechanisms against human MCF-7 cell lines	[25]
Musa paradisiaca	Peel/bract extract	Nanosilver (nAg) nanoparticles	Provides antimicrobial resistance, serves as eco-catalyst	[26]

3.4 Metallic Nanoparticles

3.4.1 *Silver Nanoparticles (Ag NPs)*

In plant-mediated synthesis, metabolites such as polyphenols and reducing sugars facilitate the rapid conversion of Ag^{+} ions under mild conditions, commonly forming spherical particles in the 10–100 nm range. The characteristics and bioactivity of these particles depend heavily on the phytochemicals involved, which act simultaneously as reducers and stabilizers.

3.4.2 *Gold Nanoparticles (Au NPs)*

Due to their exceptional stability, compatibility with biological systems, and highly evolved optical behavior [27]. In green synthesis, phytochemicals, particularly polyphenols, facilitate the reduction of Au^{3+} ions under mild conditions, while their specific chemistry governs nanoparticle morphology. Their distinct surface plasmon resonance makes them valuable for biomedical applications such as bioimaging, targeted drug delivery, and photothermal cancer therapy [28]. Furthermore, the presence of uncommon phytochemical templates in certain underutilized species can impart specialized surface and optical features, enhancing the theranostic potential of plant-derived AuNPs [29].

3.4.3 Iron-Based Nanoparticles (Fe NPs)

Magnetite (Fe_3O_4), maghemite (γ-Fe_2O_3), and zero-valent iron are frequently synthesized via plant extracts because of their intrinsic magnetic properties and versatility across multiple applications [30]. Polyphenolic compounds from plants act both as reducing agents and stabilizers, converting Fe^{3+}/Fe^{2+} ions into nanoparticles while enhancing colloidal stability. Plant extracts rich in phenolics, flavonoids, and reducing sugars have been applied to the synthesis of platinum (Pt) and palladium (Pd) nanoparticles, where these compounds function simultaneously as reducing and stabilizing agents for Pt^{2+} and Pd^{2+} ions. The resulting biogenic nanoparticles, generally smaller than 20 nm, display high catalytic activity in hydrogenation and oxidation reactions. Beyond their catalytic potential, they have also demonstrated promising antioxidant and anticancer activities, indicating their biomedical relevance [27].

3.4.4 Nickel Nanoparticles (NiNPs)

NiNPs synthesized via plant extracts are gaining interest due to their potential in catalysis, magnetism, and energy storage. Phytochemicals enable Ni^{2+} ions to be converted into stable nanoparticles under mild synthesis conditions. These green-synthesized NiNPs often display superparamagnetic properties and enhanced catalytic efficiency, which make them attractive for applications in fuel cells, sensors, and related energy technologies. The nanoparticles produced, usually smaller than 50 nm and commonly in the anatase crystalline phase, have been extensively investigated for their ability to photo-catalytically degrade organic pollutants, as well as for their antimicrobial activity and emerging biomedical applications [31].

3.5 Biomedical Applications

3.5.1 Antibacterial Activity

Nanoparticle metals, especially silver (Ag), produced using plants, are well known as broad-spectrum antibacterial agents with activity against both Gram-positive and Gram-negative microorganisms. Nanoparticles kill microbes through microbial membrane lysis and the formation of reactive oxygen species [32]. AgNPs made from *Terminalia chebula* and *Kaempferia angustifolia* are able to inhibit microorganisms in wounds and urine infection efficiently, even at low dosages [27, 29, 33].

3.5.2 Antioxidant Activity

Biogenic nanoparticles often maintain, and in certain instances even enhance, the antioxidant activity of parent botanical preparations. Such activity is particularly characteristic for phytochemicals, polyphenols, tannins, and flavonoids in particular, where on the nanoparticles' surface a stabilizing corona forms actively inhibiting free radicals. Selenium nanoparticles (SeNPs) synthesized via green approaches also demonstrate pronounced in vitro antioxidant activity [27].

3.5.3 Antidiabetic Activity

Plant-derived nanoparticles have shown remarkable promise in diabetes mellitus management. Their mechanism of action revolves mainly around inhibiting key carbohydrate-digesting enzymes, namely α-amylase and α-glucosidase, playing a major role in postprandial glucose concentration regulation in the bloodstream. For instance, copper and zinc oxide nanoparticles from *Tridax procumbens* and *Aegle marmelos* showed dose-dependent inhibitory activity for such enzymes, along with the presence of antioxidants that benefit cells responsive to insulin. Apart from such, various studies indicate a possibility of enhanced therapeutic effects through the combination of nanoparticles with plant-derived bioactive molecules, such as those encapsulated in silver/copper oxide composites. But in vivo studies on a comprehensive scale are necessary before practical use in clinics in order to explain pharmacokinetics, ion release patterns, and safety factors [27].

3.5.4 Anticancer Activity

The plant-assisted biosynthesis of metallic nanoparticles such as silver (Ag) and gold (Au) is emerging as a promising strategy for cancer therapy. Metallic nanoparticles can induce apoptosis in tumor cells by various mechanisms related to reactive oxygen species (ROS)-mediated cell death, impairment of mitochondrial viability, disruption of DNA with lysis, and enhancement of the delivery of co-administered anticancer drugs. Reports have pointed out their preference for attacking malignant cells preferentially, exemplified in cases of MCF-7 (breast) and A549 (lung) cells, though having relatively minor effects on normal cells [27, 29].

3.5.5 *Antiviral Activity*

Despite their well-documented antibacterial and anticancer effects, the antiviral effects of nanoparticles produced from plants have attracted relatively fewer studies until evidence continues to portray their efficacy. Both gold (Au) and silver (Ag) nanoparticles have the potential to retard viral infection through inhibition of viral entry, disruption of structural integrity of viral envelopes, and targeting major replication enzymes. Several studies carried out in vitro have shown inhibitory activity of AgNPs, AuNPs, and ZnO nanoparticles produced from plants against a wide array of viruses from HIV to influenza. Besides their therapeutic use, the nanoparticles have been applied in coatings on surfaces, with high virucidal effects and promising infection control [27].

3.6 Bimetallic Nanoparticles: Dual Metal Synergy

The bimetallic nanoparticles (BNPs) are formed from two different metallic elements embedded in a single nanostructure, where the metal-metal interactions generate synergistic properties absent in the single metals themselves. Reduction kinetics, mismatch in cohesive energy, mismatch in atomic size, and surface-ligand interactions influence the final BNP structure [34]. BNPs are categorized in general into three major classifications:

(a) *Alloy Nanoparticles*
At the atomic scale, whether metals mix by forming random solid solutions or ordered intermetallic compounds depends on their alloying and bonding characteristics. A good example includes Au-Ag and Au-Pd alloyed nanoparticles, which show optimized catalytic and optical properties compared to single metal nanoparticles [35].

(b) *Core-Shell Nanoparticles*
These are made up of a core made of a single metal, surrounded by a shell made of a different metal. Usually, high cohesive energy metals (Pt, Au) are the core, and low cohesive energy metals (Ag, Cu) are the shell. Examples are the Au-core/Ag-shell and Pt-core/Au-shell nanoparticles, to take advantage of the catalytic efficiency and plasmonic stability [36].

(c) *Heterostructured Nanoparticles*
This is a category of different names like Janus-type particles, separated domains, and multi-shell shapes. For example, spatially distinguished oxidation and reduction locations are found in Janus Au-Pt nanoparticles [29].

3.6.1 Ag-Cu BNPs

Ag-Cu bimetallic nanoparticles (BNPs) take advantage of the broad-spectrum antimicrobial properties of Ag^+ and the ability of Cu^{2+} to generate ROS. These BNPs can efficiently kill microorganisms living in biofilms and induce a redox imbalance in pathogens. Thirumoorthy et al. [37] expressed the promising potential of Ag-Cu via their preparation of crystalline Ag-Cu nanoparticles (7–12 nm in size), using *Aerva lanata* extract, with strong antioxidant, antibiofilm, antibacterial, and cytotoxic activities toward HeLa cells (IC_{50} = 17 µg/mL).

3.6.2 Au-Pt BNPs

Au-Pt nanoparticles combine the optical tunability and biocompatibility of Au and the catalytic activity of Pt. The incorporation of Pt alters the electronic density of Au, generating more sites for ROS formation and inducing apoptosis in cancer cells. Pt-Ag and Pt-Au BNPs prepared via phytochemical extracts are ultrasmall in size, stable, and show robust bioactivity, with their activity dependent on structural architecture [38, 39].

3.6.3 ZnO-Ag Heterostructures

ZnO-Ag composites are semiconductor-metal hybrids, where Ag forms Schottky junctions with ZnO to promote less recombination of charge carriers and greater production of reactive oxygen species under visible or UV light, thus promoting enhanced photocatalytic and antimicrobial performance. For instance, *Tetradenia riperia* extracts produced Ag-ZnO nanoparticles, resulting in good inhibition of *E. coli* [40], while *Launaea cornuta* produced Ag-ZnO of 21 nm that had inhibition zones larger than 19 mm against *E. coli* and *S. aureus* [41] (Table 3.3).

3.7 Mechanism of Action Against Human Illness

3.7.1 Disruption of Microbial Membranes

Underutilized medicinal plants act against pathogen is by disrupting microbial membranes (Fig. 3.1). Ultee et al. [57] observed that incorporation of carvacrol into nanoparticles not only disrupts the outer membrane of *E. coli* but also causes functional collapse of the cytoplasmic membrane. Moreover, apart from antibacterial effects, phytochemicals interfere with biofilm formation and quorum sensing, which

Table 3.3 Plant-mediated synthesis of nanoparticles and their applications

Plant	Nanoparticle type	Application	References
Aerva lanata	Bimetallic (Ag-Cu BNPs)	Exhibits strong antioxidant and antibiofilm activity; shows cytotoxicity against HeLa cancer cells and antibacterial effect	[37]
Launaea cornuta	Bimetallic (Ag-ZnO nanocomposites)	Demonstrates potent antibacterial activity against *E. coli* and *S. aureus*, suitable for antimicrobial coatings	[41]
Tetradenia riperia	Bimetallic (Ag-ZnO nanocomposites)	Leaf extract-mediated BNPs inhibit *E. coli* growth and show free radical scavenging (antioxidant)	[40]
Ziziphus spina-christi	Trimetallic (CuO/Ag/ZnO)	Produces nanoparticles with broad-spectrum antimicrobial and antioxidant activity, effective against resistant microbes	[42]
Ficus rumphii	Bimetallic (Zn-Cu BNPs)	Induces apoptosis in melanoma cancer cells, highlighting anticancer potential	[43]
Leucas aspera	Bimetallic (Ag-Cu BNPs)	Displays anticancer activity by inducing oxidative stress and DNA damage in cancer cells	[44]
*Azadirachta indica (*Neem)	Bimetallic (Au@Ag BNPs)	Known for enhanced antimicrobial activity due to synergistic ion release (Ag^+ and Au^+)	[45]
Swietenia mahagoni	Bimetallic (Ag/Au alloy NPs)	Exhibits dual optical/catalytic properties and antimicrobial effects	[46]
Moringa oleifera	Metallic (Au NPs)	Exhibits anticancer and antioxidant properties in green-synthesized Au NPs	[47]
Peganum harmala	Bimetallic (Ag-Cu)	Antibacterial and photocatalytic dye degradation	[48]
Terminalia chebula	Metallic (Au)	Rapid green synthesis; antimicrobial versus Gram-positive	[49]
Artemisia roxburghiana	Bimetallic (Ag-Cu)	Antioxidant and antimicrobial via green route	[50]
Curcuma longa (flower)	Metallic (Ag)	Antibacterial activity	[51]
Phyllanthus emblica	Metallic (Ag)	Antibacterial and biocompatible silver nanoparticles in vivo	[52]
Polianthes tuberosa	Metallic (Au)	Exhibits antibacterial and anticancer (MCF-7) activities	[53]
Origanum vulgare (Oregano)	Metallic (Ag)	AgNPs synthesized with strong bactericidal and fungicidal properties against various strains	[54]
Annona squamosa (Custard apple)	Metallic and bimetallic (Ag, Au, Ag-Au)	Shows antibacterial action (both Gram-positive and negative strains)	[55]
Swertia chirata	Metallic (Ag)	Antimicrobial activity against human pathogens with green synthesis	[56]

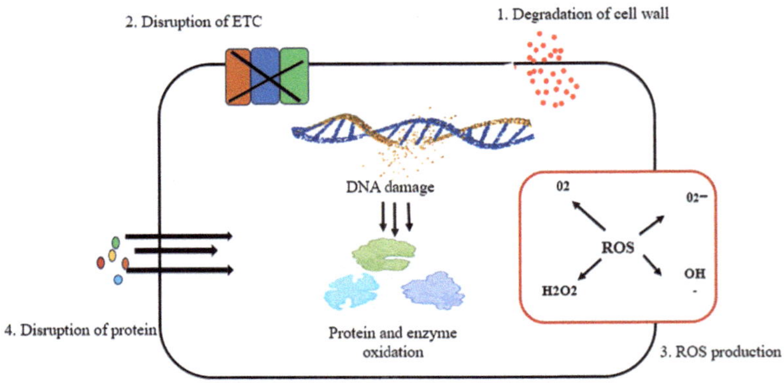

Fig. 3.1 Metallic nanoparticles showed a variety of approaches to combat human diseases. (1) Degradation of cell wall by releasing metal ions. (2) Interfere with electron transport chains and other aspects of cellular respiration. (3) Producing reactive oxygen species to induce oxidative stress, causing physical or chemical damage to cell membranes, and releasing metal ions that interfere with intracellular functions like DNA integrity and protein function. They can also change gene expression, block important metabolic pathways, and interfere with vital enzymes

are essential virulence traits. Quercetin has been reported to inhibit quorum-sensing circuits in *Pseudomonas aeruginosa* and reduce *Enterococcus faecalis* by nearly 95% [58]. This demonstrated that underutilized plants contain potent phytochemicals, and when combined with nanoparticles, their stability and therapeutic efficiency against multidrug-resistant pathogens can significantly improve. Similarly, Kumar et al. [59] observed that *Cassia tora* mediated nanoparticles induce pore formation in the membranes of *E. faecalis*, a common cause of root canal infections (Fig. 3.1).

3.7.2 Generation of Reactive Oxygen Species

Plant-based NPs are known to induce the formation of reactive oxygen species (ROS), which in excess cause cellular death. ROS affects signal transduction process pathways, which also causes cellular apoptosis [60]. Veeriah et al. [61] noted that "polyphenols may under excess amount act as pro-oxidants to initiates a ROS-mediated cellular DNA damage." For instance, curcumin-derived nanoparticles have been shown to induce ROS generation, mitochondrial membrane depolarization, and activation of caspases in tumor cells [62]. Conversely, in a normal physiological setting, many flavonoids act as antioxidants. Kumar and Pandey [63] reported that nanoparticles derived from quercetin scavenge hydroxyl and superoxide radicals and upregulate the antioxidant enzymes such as superoxide dismutase (SOD), catalase (CAT), and glutathione peroxidase (GPx).

3.7.3 *DNA/Protein Interaction*

Another important mechanism involves direct binding of nanoparticles with DNA and protein, resulting in interference with replication and cell-cycle progression. Berberine (isoquinoline alkaloid), when loaded onto nanoparticles, is capable of intercalating between DNA base pairs and destabilize double helix structure of DNA. Nanoparticles derived from *Moringa oleifera* leaves were reported to bind with bacterial DNA, which causes inhibition of replication and transcription processes [64].

3.8 Applications

3.8.1 *Cancer Treatment*

Plant-derived nanoparticles offer an eco-friendly, biocompatible, and cost-effective alternative for cancer therapy, exerting their effects by inducing apoptosis, suppressing angiogenesis, inhibiting metastasis, and modulating signaling pathways. In another study, *Kigelia africana*-mediated silver nanoparticles exhibited selective cytotoxicity against colon cancer cells while sparing normal fibroblasts, indicating their therapeutic safety margin [64]. Silver nanoparticles synthesized using *Tephrosia purpurea*, a neglected leguminous plant, were found to trigger apoptosis in human ovarian cancer cells by downregulating survivin protein expression [65].

3.8.2 *Diabetes Management*

Diabetes mellitus is a chronic metabolic condition characterized by hyperglycemia, impaired insulin signaling, oxidative stress, and delayed wound healing. Rahul et al. [66] worked on alpha amylase assay and alpha glycosidase assay (glucose tolerance test) of *T. chebula* for testing antidiabetic activity and concluded that *T. chebula* has wide antidiabetic activity due to the presence of chitosan nanoparticle. Bhavi et al. [67] worked on *Syzygium*-based AgNPs and demonstrated their ability to effectively regulate blood glucose levels, similar to conventional medications. Thus, nanoparticles from underutilized plants not only regulate blood sugar levels but also facilitate faster recovery from diabetic complications such as chronic wounds, which remain a major concern in clinical practice.

3.8.3 Neuroprotection and Inflammation

Neurological disorders such as Alzheimer's, Parkinson's, and multiple sclerosis occur due to oxidative stress, neuroinflammation, and protein aggregation. Silver nanoparticles derived from *Clerodendrum serratum* were found to downregulate pro-inflammatory cytokines (TNF-α, IL-6, and IL-1β) in lipopolysaccharide-activated macrophages, indicating their potential role in reducing neuroinflammation [68]. Similarly, gold nanoparticles prepared from *T. chebula* were shown to influence NF-κB and mitogen-activated protein kinase (MAPK) pathways in neuronal cell models, thereby preventing apoptosis caused by oxidative stress [69]. Silver nanoparticles synthesized using *Centella asiatica* leaves, a plant known for its traditionally valued neurological benefits, were demonstrated to improve memory and reduce oxidative stress in Alzheimer's disease [70].

3.9 Conclusion

Nanotechnology is an emerging branch of pharmaceutical sciences wherein the particles extend in nano sizes and turn out to be more responsive when compared with their unique counterparts. These nanoparticles (NPs) exist in a variety of forms and are categorized into different groups such as organic, inorganic, and carbon-based on the basis of their origin, properties, shape, and size. Among various biological agents used for the green synthesis, plants hold a distinct advantage because of their rich reservoir of secondary metabolites such as phenolics, flavonoids, terpenoids, alkaloids, and proteins, which act as natural reducing and stabilizing agents. Exploring underutilized wild edible plants offers a substantial solution to food security, as they are inherently rich in micronutrients, bioactive compounds, and pharmacologically significant phytochemicals. The phytochemicals present in these plants are potential reducing agents for the bio-fabrication of nanoparticles. The non-toxic nature and combination of the plant phytochemicals with precursor ions act as key aspects for the synthesis of nanoparticles. Underutilized plant-based NPs exhibited various medicinal properties against human diseases. For instances *Leucas aspera*-based bimetallic NPs (Ag-Cu) show anticancer activity by inducing oxidative stress and DNA damage in cancer cells, and (Ag-Cu) synthesized from *Artemisia roxburghiana* exhibited potent antioxidant and antimicrobial activity. The high content of carotenoids and minerals in neglected underutilized species enables them to significantly improve dietary micronutrients for millions of people. These species are frequently used by farmers as food supplements during periods of staple crop failure and emergency situations. Underutilized plants are increasingly recognized as valuable reservoirs of phytochemicals due to their adaptation to marginal environments and unique metabolic pathways.

Author's Contribution AT—Writing original draft; SS, KD, PS—Data interpretation, conceptualization; AT, SS, KD—Final formatting, final approval; MP—Data collection.

Conflicts of Interest No conflicts of interest.

Ethical Approval Not applicable.

References

1. Shrivastava S, Dash D (2009) Applying nanotechnology to human health: revolution in biomedical sciences. J Nanotechnol 1:184702. https://doi.org/10.1155/2009/184702
2. Roszek B, De Jong WH, Geertsma RE (2005) Nanotechnology in medical applications: state of the art in materials and devices. Tech Rep. 265001001 RIVM Bilthoven The Netherlands
3. Jin S, Ye K (2007) Nanoparticle mediated drug delivery and gene therapy. Biotechnol Prog 23(1):32–41. https://doi.org/10.1021/bp060348j
4. Morones JR, Elechiguerra JL, Camacho A et al (2005) The bactericidal effect of silver nanoparticles. Nano 16(10):2346. https://doi.org/10.1088/0957-4484/16/10/059
5. Nepal D, Balasubramanian S, Simonian AL et al (2008) Strong antimicrobial coatings: single-walled carbon nanotubes armored with biopolymers. Nano Lett 8(7):1896–1901
6. Dhillon GS, Brar SK, Kaur S et al (2012) Green approach for nanoparticle biosynthesis by fungi: current trends and applications. Crit Rev Biotechnol 32(1):49–73. https://doi.org/10.3109/07388551.2010.550568
7. Singh S, Vidyarthi AS, Dev A (2015) Microbial synthesis of nanoparticles: an overview. In: Singh OV (ed) Bio nanoparticles: biosynthesis and sustainable biotechnological implications, 1st edn. Wiley, Hoboken, NJ
8. Padulosi S, Mal B et al (2015) Minor crops as a central element for sustainable enhanced incomes empowerment and nutrition in rural India. J Sustain 7:8904–8933. https://doi.org/10.3390/su7078904
9. Mabhaudhi TO, Reilly P, Walker S et al (2016) Opportunities for underutilised crops in southern Africa's post–2015 development agenda. Sustain 8(4):302. https://doi.org/10.3390/su8040302
10. Padulosi S, Hoeschle ZI (2004) Underutilized plant species what are they? Leisa Magazine 20(1):5–6
11. Oxfam Novib, ANDES, CTDT, SEARICE (2016) Women seeds and nutrition. Oxfam Novib, Consolidated baseline survey report for Vietnam and Zimbabwe The Hague
12. Gajanana TM, Gowda D, Reddy BMC (2010) Reddy exploring market potential and developing linkages a case of underutilized fruit products in India. Agri Econ Res Rev 23:437–443
13. Kour S, Bakshi P, Sharma A et al (2018) Strategies on conservation improvement and utilization of underutilized fruit crops. Int J Curr Micro App Sci 7(03):638–650. https://doi.org/10.20546/ijcmas.2018.703.075
14. Scarano A, Semeraro T, Chieppa M et al (2021) Neglected and underutilized plant species (NUS) from the Apulia region worthy of being rescued and re-included in daily diet. Horticul 7(7):177. https://doi.org/10.3390/horticulturae7070177
15. Mudau FN, Chimonyo VGP et al (2022) Neglected and underutilised crops: a systematic review of their potential as food and herbal medicine crops in South Africa. Front Pharmacol 12:809–866. https://doi.org/10.3389/fphar.2021.809866
16. Talucder MSA, Ruba UB, Robi MAS (2024) Potentiality of neglected and underutilized species as a future resilient food: systematic review. J Agric Food Res 16:101–116. https://doi.org/10.1016/j.jafr.2024.101116

17. Faruq MO, Rahim A, Arifuzzaman M et al (2024) Phytochemical screening and antioxidant potential of *Amaranthus virdis* and *A. spinosus*. J Agri Food Res 18:101341. https://doi.org/10.1016/j.jafr.2024.101341
18. Vinoth K, Kumar SR (2025) Morphological molecular and pharmacological review of *Cissus quadrangularis* L. an underutilized medicinal plant. Front in Plant Sci 16:1586624. https://doi.org/10.3389/fpls.2025.1586624
19. Marappan K, Sadasivam S, Natarajan N et al (2025) Underutilized fruit crops as a sustainable approach to enhancing nutritional security and promoting economic growth. Front Sustain Food Syst 9:1618112. https://doi.org/10.3389/fsufs.2025.1618112
20. Phoswa WN, Mokgalaboni K (2023) Comprehensive overview of the effects of *Amaranthus* and *Abelmoschus esculentus* on markers of oxidative stress in diabetes mellitus. Life 13(9):1830. https://doi.org/10.3390/life13091830
21. Opris R, Tatomir C, Olteanu D et al (2017) The effect of *Sambucus nigra* L. extract and phytosinthesized gold nanoparticles on diabetic rats. Colloids Surf B: Biointerfaces 150:192–200. https://doi.org/10.1016/j.colsurfb.2016.11.033
22. Matinise N, Kaviyarasu K, Mongwaketsi N et al (2018) Green synthesis of novel zinc iron oxide (ZnFe2O4) nanocomposite via *Moringa Oleifera* natural extract for electrochemical applications. App Surf Sci 446:66–73. https://doi.org/10.1016/j.apsusc.2018.02.187
23. Lebaschi S, Hekmati M, Veisi H (2017) Green synthesis of palladium nanoparticles mediated by black tea leaves (*Camellia sinensis*) extract: catalytic activity in the reduction of 4-nitrophenol and Suzuki Miyaura coupling reaction under ligand-free conditions. J Collo Inter Sci 485:223–231. https://doi.org/10.1016/j.jcis.2016.09.027
24. Arunachalam R, Dhanasingh S et al (2012) Phytosynthesis of Silver nanoparticles using *Coccinia grandis* leaf extract and its application in the photocatalytic degradation. Collo Surf B: Biointerfaces 94:226–230. https://doi.org/10.1016/j.colsurfb.2012.01.040
25. Kabir SR, Asaduzzaman AK et al (2020) *Zizyphus mauritiana* fruit extract mediated synthesized silver/silver chloride nanoparticles retain antimicrobial activity and induce apoptosis in MCF 7 cells through the FAS pathway. ACS Omeg 5(32):20599–20608. https://doi.org/10.1021/acsomega.0c02878
26. Phan HP, Tu QD, Nguyen MT et al (2024) *Musa paradisiaca* L. peel extract bioinspired anisotropic nano silver with the multipurpose of hydrogenation eco catalyst and antimicrobial resistance. Heli 10(16)
27. Singh J, Dutta T, Kim KH et al (2018) Green synthesis of metals and their oxide nanoparticles: applications for environmental remediation. J Nanobiotechnol. https://doi.org/10.1186/s12951-018-0408-4
28. Nisha, Sachan RSK, Singh A et al (2024) Plant mediated gold nanoparticles in cancer therapy: exploring anti cancer mechanisms drug delivery applications and future prospects. Front Nanotechnol. https://doi.org/10.3389/fnano.2024.1490980
29. Berta L, Coman NA, Rusu A, Tanase C (2021) A review on plant-mediated synthesis of bimetallic nanoparticles, characterisation and their biological applications. Dent Mater 14(24):7677. https://doi.org/10.3390/ma14247677
30. Elkhateeb O, Atta MB, Mahmoud E (2024) Biosynthesis of iron oxide nanoparticles using plant extracts and evaluation of their antibacterial activity. AMB Express 14(1):92. https://doi.org/10.1186/s13568-024-01746-9
31. Fierascu RC, Fierascu I, Ortan A et al (2021) *Plantago media* L.-Explored and potential applications of an underutilized plant. Plan 10(2):265. https://doi.org/10.3390/plants10020265
32. More PR, Pandit S, Filippis AD et al (2023) Silver nanoparticles: bactericidal and mechanistic approach against drug resistant pathogens. Microorganisms 11(2):369. https://doi.org/10.3390/microorganisms11020369
33. Chakraborty A, Haque SM, Ghosh D et al (2022) Phytogenic silver nanoparticles from tissue cultured *Kaempferia angustifolia* underutilized medicinal herb: a comparative antibacterial study on urinary pathogens. J Genet Eng Biotechnol 20:131. https://doi.org/10.1186/s43141-022-00414-4

34. Singh C, Mehata AK, Priya V et al (2022) Bimetallic Au–Ag nanoparticles: advanced nanotechnology for tackling antimicrobial resistance. Molecules 27(20):7059. https://doi.org/10.3390/molecules27207059
35. Wang Y, Shi H, Zhao D et al (2024) Lattice-strained bimetallic nanocatalysts: fundamentals of synthesis and structure. Molecules 29(13):3062. https://doi.org/10.3390/molecules29133062
36. Eom N, Messing ME, Johansson J et al (2021) General trends in core–shell preferences for bimetallic nanoparticles. ACS Nano 15(5):8883–8895. https://doi.org/10.1021/acsnano.1c01500
37. Thirumoorthy G, Balasubramanian B, George JA et al (2024) Phytofabricated bimetallic synthesis of silver-copper nanoparticles using *Aerva lanata* extract to evaluate their potential cytotoxic and antimicrobial activities. Sci Rep 14(1):1270. https://doi.org/10.1038/s41598-024-51647-x
38. Bekmezci M, Ozturk H, Akin M et al (2023) Bimetallic biogenic Pt-Ag nanoparticle and their application for electrochemical dopamine sensor. Biosen 13(5):531. https://doi.org/10.3390/bios13050531
39. Nagarajan SB, Ramakrishnan S, Jayaraman A (2024) Theranostic aspects of palladium-based bimetallic nanoparticles in biomedical field: a state-of-the-art. Health Care Sci 3(3):181–202. https://doi.org/10.1002/hcs2.96
40. Mtavangu SG, Machunda RL, Van der BB et al (2022) In situ facile green synthesis of Ag–ZnO nanocomposites using *Tetradenia riperia* leaf extract and its antimicrobial efficacy on water disinfection. Sci Rep 12(1):15359. https://doi.org/10.1038/s41598-022-19403-1
41. Makauki E, Mtavangu SG, Basu OD et al (2023) Facile biosynthesis of Ag–ZnO nanocomposites using *Launaea cornuta* leaf extract and their antimicrobial activity. Discover Nano 18(1):142. https://doi.org/10.1186/s11671-023-03925-2
42. Afzal MA, Javed M, Aroob S et al (2023) The biogenic synthesis of bimetallic Ag/ZnO nanoparticles: a multifunctional approach for methyl violet photocatalytic degradation and the assessment of antibacterial, antioxidant, and cytotoxicity properties. Nanomat 13(14):2079. https://doi.org/10.3390/nano13142079
43. Cao Y, Dhahad HA, El-Shorbagy MA et al (2021) Green synthesis of bimetallic ZnO–CuO nanoparticles and their cytotoxicity properties. Sci Rep 11(1):23479. https://doi.org/10.1038/s41598-021-02937-1
44. Idris DS, Roy A (2023) Synthesis of bimetallic nanoparticles and applications-an updated review. Crystals 13(4):637. https://doi.org/10.3390/cryst13040637
45. Ahmed S, Saifullah, Ahmad M et al (2017) Green synthesis of silver nanoparticles using *Clerodendrum inerme* and their antimicrobial activities. Mater Lett 186:122–125
46. Chamakura K, Perez-Ballestero R, Luo Z et al (2011) Comparison of bacterial activities of silver nanoparticles with common chemical disinfectants. Coll Surf B: Biointerfaces 84(1):88–96. https://doi.org/10.1016/j.colsurfb.2010.12.020
47. Duarte M, Kuchibhatla M, Khandelwal et al (2019) Heterogeneity in neutrophil responses to immune complexes. Blood Advan 3(19):2778–2789. https://doi.org/10.1182/bloodadvances.2019000235
48. Malik MA, Albeladi SS, Al-Maaqar SM et al (2023) Biosynthesis of novel Ag Cu bimetallic nanoparticles from leaf extract of *Salvia officinalis* and their antibacterial activity. Life 13(3):653. https://doi.org/10.3390/life13030653
49. Kumar KM, Sinha M, Mandal BK et al (2012) Green synthesis of silver nanoparticles using *Terminalia chebula* extract at room temperature and their antimicrobial studies. Spectrochim Acta A Mol Biomol Spectrosc 91:228–233. https://doi.org/10.1016/j.saa.2012.02.001
50. Tripathi PK, Sati SC (2019) Green synthesis of Ag Cu bimetallic nanoparticles and determination of its antimicrobial and antioxidant activities. J Emerg Technol Innov Res 6(4):627–633
51. Rajak KK, Pahilani P, Patel H et al (2023) Green synthesis of silver nanoparticles using *Curcuma longa* flower extract and antibacterial activity. arXiv:2304.04777. https://doi.org/10.48550/arXiv.2304.04777

52. Hossain MM, Hamza A, Polash SK et al (2024) Green synthesis of silver nanoparticles using *Phyllanthus emblica* extract: investigation of antibacterial activity and biocompatibility in vivo. RSC Pharm 1(2):245–258. https://doi.org/10.1039/D3PM00077J
53. Alghuthaymi MA, Rajkuberan C, Santhiya T et al (2021) Green synthesis of gold nanoparticles using *Polianthes tuberosa* L. floral extract. Plants 10(11):2370. https://doi.org/10.3390/plants10112370
54. Shaik MR, Khan M, Kuniyil M et al (2018) Plant extract assisted green synthesis of silver nanoparticles using *Origanum vulgare* L. extract and their microbicidal activities. Sustain 10(4):913. https://doi.org/10.3390/su10040913
55. Jibrin F, Fanoro OT, Maluleke R et al (2024) Biosynthesis characterization and antibacterial activity of gold silver and bimetallic nanoparticles using *Annona squamosa* L. leaves. Antibiotics 13(12):1199. https://doi.org/10.3390/antibiotics13121199
56. Shereen MA, Ahmad A, Khan H et al (2024) Plant extract preparation and green synthesis of silver nanoparticles using *Swertia chirata*: characterization and antimicrobial activity against selected human pathogens. Heliyon 10(6). https://doi.org/10.1016/j.heliyon.2024.e28038
57. Ultee A, Bennik MH, Moezelaar R (2002) The phenolic hydroxyl group of carvacrol is essential for action against the food borne pathogen *Bacillus cereus*. App Environ Microbiol 68(4):1561–1568. https://doi.org/10.1128/AEM.68.4.1561-1568.2002
58. Hawser S, Islam K, Li C (2021) Quercetin inhibits quorum sensing and biofilm formation in *Enterococcus faecalis*. J Glob Antimicro Resist 25:123–131
59. Kumar V, Sharma A, Singh A (2019) Antibacterial mechanism of silver nanoparticles synthesized from *Cassia tora* leaves against multidrug resistant pathogens. Mat Sci Eng 98:105467
60. Ovais M, Khalil AT, Raza A et al (2016) Green synthesis of silver nanoparticles via plant extracts: beginning a new era in cancer theranostics. Nanomedicine 1:3157–3177. https://doi.org/10.2217/nnm-2016-0279
61. Veeriah S, Kautenburger T, Habermann N et al (2015) Polyphenols modulate oxidative stress in cancer cells: potential dual role. Oxidat Med Cell Longe 2015:713
62. Agarwal A, Sharma V, Arora R, Gupta M (2018) Curcumin induces ROS-mediated apoptosis and cell cycle arrest in HT29 cells. Nut Res 51:67–81
63. Kumar S, Pandey AK (2017) Chemistry and biological activities of flavonoids: an overview. Sci World J 2017:162750. https://doi.org/10.1155/2013/162750
64. Sharma D, Kanchi S, Bisetty K (2019) Biogenic synthesis of nanoparticles: a review. Arab J Chem 12(8):3576–3600. https://doi.org/10.1016/j.arabjc.2015.11.002
65. Shashidhara V, Srinivasappa PM, Jadhav AH et al (2023) Phytosynthesis of silver nanoparticles from Tephrosia purpurea and its in vitro biogenic actions. App Organomet Chem:1–12. https://doi.org/10.1002/aoc.7266
66. Nath R, Rajakumari K, Thiruchelvi R et al (2023) Chitosan nanoparticle encapsulated with *Terminalia chebula* in vitro antidiabetic activity. Oxid Comm 46(1):139
67. Bhavi SM, Abbigeri MB, Thokchom B et al (2024) *Syzygium* genus: exploring plant based synthesis of silver nanoparticles for antidiabetic efficacy. Imp Biotech:279–293
68. Veerasamy R, Xin TZ, Gunasagaran S et al (2011) Anti inflammatory potential of silver nanoparticles synthesized from *Clerodendrum serratum*. Coll Surf B: Biointer 83(1):16–22. https://doi.org/10.1016/j.jscs.2010.06.004
69. Ramesh PS, Kokila T, Geetha D (2015) Plant mediated green synthesis and antibacterial activity of silver nanoparticles using *Terminalia chebula* fruit extract. Spectrochim Acta A Mol Biomol Spectr 142:339–343. https://doi.org/10.1016/j.saa.2015.01.062
70. Lokanathan Y, Omar N, Ahmed Puzi NN et al (2016) Recent updates in neuroprotective and neuroregenerative potential of *Centella asiatica*. Malay J Med Sci 23(1):4–14

Chapter 4
Role of Underutilized Plant-Derived Nanoparticles in Combating Food-Borne Bacteria

Junaid Aman, Radha Khirwar, and Maryam Iqbal

Abstract *Salmonella enterica*, *Escherichia coli* O157:H7, and *Listeria monocytogenes* are food-borne pathogens, which are common in causing disease and resulting in economic losses in many parts of the world. Traditional preservatives such as sodium benzoate, potassium sorbate, sodium nitrite, and sulfur dioxide, along with antibiotics, have been restricted by antimicrobial resistance and the issue of chemical additives by consumers. Green synthesis is emerging as an eco-friendly alternative to the use of nanoparticles (NPs) or nanoparticles derived from plants. Phytochemicals with reducing, stabilizing, and antimicrobial properties are abundant in underutilized plant species such as Jamun (*Syzygium cumini*), mango (*Mangifera indica*), pomegranate (*Punica granatum*), banana (*Musa paradisiaca*), tamarind (*Tamarindus indica*), and orange (*Citrus sinensis*) derived from peels, seeds, and agro-waste. Membrane disruption, the formation of reactive oxygen species, enzyme inactivation, and the prevention of biofilms are released by these NPs. Some of its uses are active packaging, edible coatings, sanitizing treatments, and nanoencapsulation, which enhance food safety, shelf life, and waste reduction. Plant-based nanotechnology is a sustainable and low-cost approach that can be used in light of the objectives of circular bioeconomy, despite difficulties in standardization, safety, and regulation.

Keywords Food-borne pathogens · plant-derived nanoparticles · antimicrobial resistance · green synthesis · biofilm inhibition · active packaging

J. Aman (✉) · R. Khirwar
School of Applied and Life Sciences, Uttaranchal University, Dehradun, India

M. Iqbal
Department of wildlife and Fisheries, University of Agriculture, Faisalabad, Pakistan

© The Author(s), under exclusive license to Springer Nature Switzerland AG 2026

E. Nepovimova et al. (eds.), *Underutilized Plants Based-Nanoparticles: Herbal Remedy to Treat Human Diseases*, https://doi.org/10.1007/978-3-032-20384-7_4

Abbreviations

AgNPs	Silver Nanoparticles
AMR	Antimicrobial Resistance
CaONPs	Calcium Oxide Nanoparticles
CFU	Colony Forming Units
CuNPs	Copper Nanoparticles
DNA	Deoxyribonucleic Acid
E. coli	*Escherichia coli*
FAO	Food and Agriculture Organization
GRAS	Generally Recognized As Safe
MRSA	Methicillin-Resistant *Staphylococcus aureus*
NPs	Nanoparticles
pH	Potential of Hydrogen
ROS	Reactive Oxygen Species
S. aureus	*Staphylococcus aureus*
Salmonella spp.	*Salmonella* species
WHO	World Health Organization
ZnO NPs	Zinc Oxide Nanoparticles

4.1 Introduction

Food-borne pathogens have been a worldwide issue of global concern, as they cause millions of diseases and high death rates annually. *Salmonella enterica*, *Escherichia coli O157:H7*, *Listeria monocytogenes*, *Staphylococcus aureus*, and *Bacillus cereus* are especially dangerous because they are widespread not only in fresh produce, meat, dairy, and processed food but also are common in food. As stated by the World Health Organization, unsafe food causes over 600 million diseases and 420,000 deaths each year, which proves the extreme necessity of effective food safety measures. Besides the direct health hazards, food-borne infections also have high economic costs in terms of food recall, health care expenses, and trade embargo. Traditional food preservation does use a great deal of chemical preservatives (sorbates, benzoates, nitrites) and antibiotics to control microbial growth. Although these approaches have worked, their shortcomings are coming out prominently. The emergence of antimicrobial resistance (AMR), consumer anxieties over chemical additives, and the possibility of toxicological consequences of synthetic preservatives have created a need for safer and more sustainable antimicrobial options. Such restrictions have led scientists to consider natural bioactive compounds and new methods that can increase the quality and safety of foods without affecting their acceptance by consumers [1].

Nanotechnology is one of the promising methods in the food industry. Nanoparticles (NPs) have distinctive physicochemical characteristics including high surface/volume ratio, controllable surface chemistry, and amplified reactivity, which allow them to bind effectively with microbial cells. Plant-derived nanoparticles are currently being investigated as a method to prevent microbial growth, biofilm formation, and as carriers in the release of natural antimicrobials in a monitored manner in food safety. In addition, the addition of nanoparticles in active packaging films, edible coatings, and nanoencapsulation has demonstrated potential in increasing shelf-life, preserving food quality, and reducing food waste. In this context, the underutilized plants are a good resource that is not fully exploited in the production of nanoparticles [2].

These comprise wild edibles, farm rejects, fruit by-products, and seeds, as well as agro-waste substances that are abundant in polyphenols, flavonoids, terpenoids, alkaloids, and polysaccharides. These phytochemicals are not only effective as reducing and stabilizing agents in the green synthesis of nanoparticles, but also provide extra antimicrobial action. The ability to use underutilized plants does not only valorize (enhancing the value of low-cost or waste materials by converting them into useful, high-value products) agricultural wastes but also contributes to low-cost, environmentally friendly, and sustainable nanotechnology development, which is consistent with the principles of a circular bioeconomy [3]. Food-borne bacterial contamination remains one of the biggest dilemmas that affects food security in the world, especially with the increased complexity and intermingling of the modern food supply chains. Increasing transportation of fresh products, processed food, and ready-to-eat food over long distances has increased the chances of transmitting microbes, causing the rapid spread of pathogens between the farm and the final consumer. Epidemic outbreaks associated with such pathogens as *Salmonella*, *E. coli*, and *Listeria* underscore the constraints of the conventional approach to sanitation and preservation, which in most cases does not offer any long-term or universal coverage. Even though thermal processing, chemical disinfectants, and traditional preservatives have been employed in the past in preserving food quality, their usefulness is declining in line with upward movements in microbial resistance, including the issue of toxicity, sensory degradation, and legislative prohibitions. The details of the labeling that are desired to be cleaner along with the minimally processed foods, the process of seeking innovative, natural, and effective antimicrobial intervention has become particularly important in the search [4].

Within the past few years, nanotechnology has come out as the paradigm platform in food safety investigations owing to the outstanding functionality displayed at the nanoscale. Nanoparticles have been shown to be highly antimicrobial due to their small size that enables them to enter bacterial cell membranes, produce reactive oxygen species (ROS), and interfere with the normal functioning of cells at low levels. Although there has been a great potential demonstrated by metallic nanoparticles such as silver, zinc oxide, and copper oxide, the associated environmental and health effects linked to chemically synthesized nanoparticles have ignited a change towards the more environmentally friendly options. Plant nanoparticles manufactured via biologically friendly methods are an eco-friendly sustainable solution as they use plant metabolites as reducing, stabilizing, and capping methodologies. These phytochemicals include flavonoid acids and phenyl acids (polyphenols) and terpenoids and alkaloids that not only aid in the formation of nanoparticles, but also interact synergistically to improve antimicrobial action, as well [5].

Of particular interest have been underutilized plants and agricultural by-products, which are inexpensive, have a large volume, and are renewable sources of nanoparticles. Numerous of these materials, such as fruit peels, seeds, leaves, and pomace, are usually discarded even though they contain lots of bioactive compounds that have the ability to initiate the creation of nanoparticles. The logical value of converting these abandoned resources to useful nanomaterials is twofold: it can help to solve the food safety problem at the same time as help to valorize the waste of agronomy and contribute to the sustainability of the environment. Additionally, the nanoparticles of these plants can also serve in numerous food protection programs, i.e., antimicrobial packaging, edible coating, surface sanitizing solution, and systems of nanoencapsulation, and thus they may be used as a wide-range relief to the whole vast field of food production. As there is an immediate necessity to develop safer and environmentally friendly antimicrobial nanoparticles technologies, and a growing concern among antimicrobial technologies with the principles of a circular bioeconomy, underutilized plant-based nanoparticles are an open door in antimicrobial technologies against bacteria contaminating food. Their multitasking feature, together with the possibility of low-cost production and its scalability, makes them lucrative objects of next-generation food safety interventions [6].

4.2 Nanoparticle Synthesis Using Underutilized Plant

Plants that are underutilized and are often ignored in mainstream agriculture and food production include a variety of species and their derivatives including fruit seeds, fruit peels, leaves, and pomace. All these natural resources are rich in bioactive compounds such as polyphenols, flavonoids, tannins, alkaloids, and terpenoids that can be used as natural reducing and stabilizing agents in the synthesis of nanoparticles with good antimicrobial potential, thus a sustainable source of green nanomaterials. The sources, which include underutilized plant foundations and their agro-industrial wastes, have demonstrated a tremendous potential in the development of nanoparticles (NPs) with a strong inhibitory effect against a wide spectrum of food-borne bacteria. As an example, *Syzygium cumini* seeds (jamun) that contain flavonoids, tannins, ellagic acid, and anthocyanins have been used to synthesize carbon dots and AgNP (AgNP) greenly, which can be used to combat *E. coli*, *Salmonella typhimurium*, and *S. aureus* because they damage cell walls and augment oxidative stress [5, 7].

Likewise, pomegranate (*Punica granatum*) peel extracts containing a significant amount of ellagitannins and phenolic acids have been utilized to produce AgNPs with a high level of bactericidal properties against *L. monocytogenes*, *E. coli*, and *Bacillus cereus* by disrupting cell membrane integrity and releasing cytoplasmic content. The leaf-derived AgNPs based on guava (*Psidium guajava*) leaves with a high concentration of flavonoids and polyphenols have a substantial inhibition zone against *S. enterica*, *E. coli*, and *Pseudomonas aeruginosa*, which is mostly due to the generation of reactive oxygen species (ROS) and the disruption of bacterial DNA replication [2]. Similarly, AgNPs and CuNPs prepared by using *Aegle marmelos* (bael) leaves and fruits, which contain alkaloids and terpenoids, have a strong antimicrobial effect against *Salmonella*, *Pseudomonas*, and *E. coli* due to their capacity to destabilize cell membrane potential and protein synthesis [8]. Moreover, citrus pomace (source of pectin, limonene, and phenolic acids) has been utilized to make nanogels and nanoemulsions that inhibit the growth of microbes on food-contact surfaces and perishable foods by preventing bacterial adhesion and biofilm formation [9]. Additionally, the seeds of *Annona squanosa* (custard apple), which are rich in *acetogenins and phenolics*, have been utilized in the synthesis of calcium oxide nanoparticles (CaONPs) with excellent bactericidal effect on *E. coli and S. aureus* by causing oxidative stress and cell lysis [10].

Taken together, all these results indicate that unused plants can be used as a cheap, environmentally friendly, and renewable bioreactor to produce nanoparticles to provide an efficient and sustainable approach to the problem of food-borne pathogens. The utilization of plants as biofactories to produce nanoparticles has received a phenomenal scientific focus, considering the growing production demand of safe, cost-effective, and sustainable antimicrobial substances in the food industry. Plant-mediated nanoparticle production differs in that, unlike other traditional methods of synthesis (chemical and physical), the conditions required are often high temperatures, toxic solvents, energy-consuming reactions (among many other undesirable conditions), with plant-mediated nanoparticle synthesis being a green technology that uses the inherent biochemical repertoire of the botanical tissue to produce nanoparticle assemblies. A large number of plant species, particularly those that fall under the underutilized or neglected category in commercial agriculture, have a large catalogue of secondary metabolites that are able to reduce metal ions to stable nanoscale. These metabolites are phenolic acids, reducing sugars, essential oils, proteins, as well as organic acids that not only begin nanoparticle formation, but also cap and stabilize these particles, which influence their size, morphology, and functioning characteristics [11].

An underutilized botanical also presents an exceptionally useful platform since they are often the waste materials that are produced during the processing of fruits, herbs extraction and harvesting of agricultural farms. Seed coats, rhizomes, bark, pulps, fibrous residues, and other materials are also, in many cases, unused in spite of their high concentrations of bioactive phytochemicals. The utilization of these materials to produce nanoparticles not only reduces the burden on the environment but also converts low-value biomass into excellent antimicrobial functional materials. Due to the increased demand for circular bioeconomy models, conversion of the plant waste into nanomaterials is a strategic pathway to enhance the sustainability in the food processing and packaging systems. The green synthesis through plants can also be further advantageous because it has the inherent ability to be made biocompatible and scaled. [6] says that, in comparison to microbial or enzyme synthesis, which involves sterile fermentation conditions, preparation, storage, and utilization of plant extracts can be achieved in large amounts in a matter of hours to become a standardized reaction substrate. This enables reproducible manufacturing of the metallic as well as non-metallic nanoparticles with linear antimicrobial diagramming to various food-borne pathogens. Besides, the structural variability of phytochemicals in underutilized plant resources results in the formation of nanoparticles with different surface functionalities, allowing the targeted action of bacteria membranes, biofilm matrix, and intracellular components. The mild conditions of the synthesis do not deactivate numerous natural antimicrobial substances of the plant, which form synergistic nanoparticle-phytochemical hybrids with better bactericidal properties in comparison with chemically synthesized analogs. Plant-derived nanoparticles are providing this wide platform as the food industry shifts to greener preservation mechanisms and greener antimicrobial solutions that are natural and residue-free and are reshaping the modern world of consumer attitudes. Their possible uses are not confined to packaging and edible coating to decontamination treatment and controlled release systems [12].

Therefore, plants that have yet to be exploited can be seen as a prospective, renewable, but cost-effective raw material in future nanotechnology applications against food-borne bacteria in a sustainable setup. Table 4.1 lists examples of plant-derived nanoparticles and the phytochemical compositions of those.

Table 4.1 Examples of underutilized plants for nanoparticle synthesis, phytochemicals, and food-borne bacterial inhibition [13, 14]

Underutilized plant [scientific name]	Type of nanoparticle [NP]	Major phytochemicals involved	Food-borne bacteria inhibited	Mechanism/mode of action
Syzygium cumini [Jamun] seeds	Carbon dots, AgNPs	Flavonoids, tannins, ellagic acid, anthocyanins	*E. coli*, *Salmonella typhimurium*, *S. aureus*	Disruption of cell wall, oxidative stress, leakage of intracellular contents
Punica granatum [Pomegranate] peel	AgNPs	Ellagitannins, phenolics, flavonoids	*Listeria monocytogenes*, *E. coli*, *Bacillus cereus*	Membrane damage, protein denaturation, cytoplasmic leakage
Psidium guajava [Guava] leaves	AgNPs	Flavonoids, polyphenols, terpenoids	*Salmonella enterica*, *E. coli*, *Pseudomonas aeruginosa*	ROS generation, inhibition of DNA replication, cell membrane disruption
Aegle marmelos [Bael] leaves and fruits	AgNPs, CuNPs	Alkaloids, terpenoids, coumarins	*Salmonella* spp., *Pseudomonas* spp., *E. coli*	Disruption of cell membrane potential, inhibition of protein synthesis
Citrus pomace [from *Citrus sinensis* and related species]	Nanogels, nanoemulsions	Pectin, limonene, phenolic acids	*E. coli*, *Listeria monocytogenes*, *S. Aureus*	Prevention of bacterial adhesion, biofilm inhibition on food-contact surfaces

The table has provided a clear picture of the under-utilized vegetation and the functions of preparing nanoparticles to counter bacteria causing food-borne diseases. *A. marmelos* (Bael), Citrus pomace, jamun (*S. cumini*), *P. granatum*, AgNPs, copper nanoparticles (CuNPs), nanogels, and nanoemulsions are all connected to particular types of nanoparticles (carbon dots, AgNPs, copper nanoparticles CuNPs, nanogels, and nanoemulsions) [13]. The plant extracts containing flavonoids, tannins, ellagic acid, anthocyanins, ellagitannins, phenolics, alkaloids, terpenoids, coumarins, pectin, limonene, and phenolic acids are used to get these nanoparticles, which provide them with antibacterial characteristics [15, 16].

These phytochemicals have a wide range of food-borne pathogens, namely, *E. coli*, *Salmonella* spp., *S. aureus*, *L. monocytogenes*, *B. cereus*, *P. aeruginosa*, and much more [17]. The mechanism of action is different: Jamun seed nanoparticles cause damage to the bacterial cell wall and oxidative stress, as a result of which intracellular material is leaked. Nanoparticles of pomegranate peel result in membrane damage, denaturation of proteins, and leakage of the cytoplasm [18]. The nanoparticles of guava leaves produce ROS, prevent the replication of DNA, and destroy the cell membrane. Bael nanoparticles affect the cell membrane potential and protein synthesis. Citrus pomace nanoparticles inhibit the growth of biomaterials on food-contact surfaces and avert bacterial adhesion [7].

These findings are important because it is possible to create plant-based antimicrobial agents in place of synthetic antibiotics that are eco-friendly. It can be used as a sustainable method of food safety by employing nanoparticles made of these underutilized plants that do not lead to the damage of beneficial microorganisms but rather harmful bacteria. The table highlights the versatility of these plant-derived nanoparticles and their applicability in addressing the growing issue of antibiotic resistance and the craving to resort to natural preserving food substances to maintain safety [5].

In addition to the particular cases that have been mentioned in the table, what stands out is the overall potential of the largely underutilized botanical resources to act as multifunctional nanomaterial platforms. Most of these plant species are inherently programmed to live in adverse environmental conditions, and this normally leads to the accumulation of highly reactive secondary metabolites having high reducing and antioxidant activities. In cases where these metabolites are employed in the formation of nanoparticles, they not only determine the physicochemical properties of the nanostructures produced, but they also increase their antimicrobial activity by synergy. Further, the diversity of phytochemicals in the form of nanoparticles with different shapes, surface charging, and functional group considerations associated with their effect on bacterial membranes and interior elements, can be contrasted, due to the structural diversity of phytochemicals within these neglected plants. It would be of special use with food-borne pathogens with vastly different cell wall structures, which include Gram-negative *E. coli* and *Salmonella* compared to Gram-positive *Listeria* and *S. aureus*. Furthermore, since plant-based nanomaterials can be genetically engineered to meet the requirements of every relevant food matrix or contamination, they represent versatile candidates in addition to being added to smart packaging, active coating, and antisepsis. They are also biodegradable and lowly toxic, which makes them even more suitable for their application in food-processing, which is in line with the existing worldwide demands to limit the use of synthetic chemicals and minimize environmental harm. The evidence as a whole supports the notion that underutilized plants are a poorly tapped but rich source of producible nanostructures that would yield both generalized and specific elimination of dangerous microbes that contaminate food [19].

4.3 Antibacterial Mechanisms of Underutilized Plant-Derived Nanoparticles

There is a broad-spectrum antibacterial effect of plant-derived nanoparticles (NPs) through multiple mechanisms that interfere with the normal activities of the pathogens of food products. The direct contact with bacterial membranes is considered to be one of the most prevalent action mechanisms during which NPs combine with the negatively charged surface of cells, and structural damage, permeability, and cytoplasmic contents leakage take place. Such damage in most instances is further complemented by the generation of reactive oxygen species (ROS) in the form of hydroxyl radicals and superoxide anions and leads to oxidative stress and disruption of redox balance in the cell [5]. ROS-mediated toxicity. Denaturation or oxidation of sulfhydryl groups leading to enzymatic and protein inactivation of critical metabolic pathways [20]. Moreover, nanoparticles may enter the bacterial cells and get bound with the DNA and cause strand breaks and mutations, as well as inhibit the replication and transcription mechanisms [21]. In addition to these deadly effects, plant-based NPs disrupt bacterial communication through the inhibition of quorum sensing, an essential signaling pathway of virulence and biofilm formation. The interference decreases the toxin synthesis and disrupts the bacterial adaptability within the food systems [22]. Nanoparticles are also highly effective in antibiofilm by inhibiting initial microbial adhesion, destabilizing existing biofilm matrices, and increasing the ability of antimicrobial agent penetration. [23] says, AgNP decreased the bacterial attachment and biofilm formation of *P. aeruginosa* and other strains; the authors note that the AgNP exposure destabilized the biofilm matrix and made the biofilm cell itself more susceptible to antibiotics, which is in line with the increased drug penetration being reported on biofilm cells destabilized by the AgNP exposure [24] stated that Ag- and Cu-nanoparticles produced by bio-synthesis were assessed in bacterial attachment effects as well as cell viability within mature biofilms. Reduced adhesion of the microbes and the viability within the biofilm were recorded in the paper, suggesting inhibition of the adhesion effect as well as destabilizing and penetrating effects. Kin Notably, mentioned synergistic interaction between green-synthesized NPs and traditional antibiotics has been noted in a number of studies, with nanoparticles increasing antibiotic absorption, reducing effective doses, and overcoming multidrug resistance in food-borne bacteria [25]. All these various mechanisms combine to highlight the possibility of plant-derived nanoparticles, which are underutilized as a promising and efficient approach in the control of food-borne bacteria. Proposed antibacterial mechanisms of plant-derived nanoparticles against food-borne pathogens [26, 27].

4.4 Applications in Food Safety

Plant-based nanoparticles (NPs) of underexploited resources have become a promising tool in food protection owing to their antimicrobial, antioxidant, and biocompatible potential. Active packaging films that include nanoparticles like AgNPs or ZnO NPs can be produced by lengthening the shelf life of perishable products because they regulate microbial growth and oxidation [28]. Plant-based nanoparticles, when used to reinforce an edible coating of biopolymers such as starch or pectin, form a natural barrier that keeps foods stable and reduces the occurrence of pathogen contamination. Nanoparticle solutions based on citrus or pomegranate waste have the potential to inactivate foodborne bacteria. The nanoparticles deactivate foodborne bacteria by releasing reactive oxygen species and bioactive ions which destroy cell membranes, interfere with metabolism, and inhibit bacteria's adhesion to food and food contact surfaces. Moreover, nano-encapsulated plant extracts increase the stability of antimicrobial phytochemicals and their release into food systems, and this improves their efficacy [29].

In addition to antimicrobial properties, nanoparticles produced out of underutilized plant ingredients are becoming macromolecules multiplied in functionality as biosensors to enhance overall food quality and safety. Intelligent packaging systems can use these plant-based nanomaterials, which will serve as food preservatives as well as food spoilage detectors. As an example, nanoparticles that contain phytochemicals will be capable of reacting to pH changes or the liberation of volatile substances by spoiled food to watch the food in real-time without chemicals. This dual functionality goes a long way in ensuring that the packaged foods are much safer with less waste [3].

Nanoparticle-impregnated biodegradable polymers are better in providing mechanical strength, water resistance, barrier to oxygen, and barrier to water in food preservation. This kind of improvement reduces the rate of moisture movement and lipid permeation which are two major causes of quality deterioration in dairy products, meat, and bakery products. Since most underutilized plants are good sources of tannins, polyphenols, or terpenoids, the nanoparticles of those plants add direct protection against rancidity and enzymatic spoilage [1].

Also, nanoemulsions and nanogels of plant products, which have been made using fruit and seed by-products, have been established to have high capacity as surface sanitizing agents in the food processing environment. The nanosystems have been used to evenly distribute antimicrobial compounds on the cutting boards, conveyor belts, and stainless-steel equipment at low concentrations compared to the crude plant extracts, and they are found to be more microbiocidal. Their nano size enhances contact efficiency with the bacteria cells, hence being very effective against biofilm-forming pathogens that are very prevalent in food facilities [30].

The second new use that has emerged is the nanoencapsulation of essential oils and phenolic compounds which ensures that sensitive bioactive molecules are not destroyed by heat, light, or changes in pH. These encapsulated agents give slow and prolonged release of antimicrobial agents when incorporated into food matrices, thereby maintaining long-term microbial repression without any sensory modifications. This is particularly important for minimally processed foods, where maintaining natural texture and flavor is essential. Overall, the integration of plant-derived nanoparticles into food systems supports clean-label preservation strategies, reduces dependency on synthetic additives, and promotes the utilization of agricultural by-products. Aligning with modern sustainability goals in the food industry [27]. Role of plant-derived nanoparticles, example materials/NPs, benefits in food safety, in all application areas shown in Table 4.2.

Nanoparticle plants are finding application in various fields of food safety because of their excellent antimicrobial properties, antioxidants, and sensing properties. AgNP extracted by pomegranate peel and citrus waste are added in active packaging systems in which films made of biopolymer are impregnated to inhibit the growth of the microbes and effectively prolong the shelf life of the food packed. Zinc oxide nanoparticles used to design a highly protective layer on food surfaces, combined with natural polymers such as starch or pectin gives an extra layer of protection to food material, further protecting their freshness and minimizing the chances of food contamination. Nanoparticle-based sanitizing treatments (such as preparing solutions using citrus peel-produced AgNP or guava leaf nanoparticles) are now used to wash fresh produce and surfaces in contact with food and significantly reduce the load of pathogens on foods and food-contact surfaces, reducing the quantities of *E. coli*, *Salmonella*, and *Listeria*. The stability, delivery, and

Table 4.2 Applications of underutilized plant-derived nanoparticles in food safety [27, 31]

Application area	Role of plant-derived nanoparticles	Example materials/ NPs	Benefits in food safety
Active packaging films	Incorporated into biopolymer films to inhibit microbial growth	AgNPs (pomegranate peel, citrus waste)	Extends shelf life, prevents spoilage
Edible coatings	Coatings with NPs form protective antimicrobial and antioxidant barriers	ZnO NPs with starch/pectin	Maintains freshness, prevents contamination
Food sanitizing treatments	NP solutions for washing fresh produce and surfaces	Citrus peel-AgNPs, guava leaf NPs	Reduces pathogens (*E. coli*, *Salmonella*, *Listeria*)
Nanoencapsulation	Encapsulation of plant extracts to enhance delivery and stability	Polymeric NPs from starch, cellulose	Improves bioavailability and controlled release
Smart packaging and biosensors	NPs integrated into sensors for detection of pathogens/toxins	Carbon dots, AgNP-based biosensors	Real-time monitoring of food contamination and quality

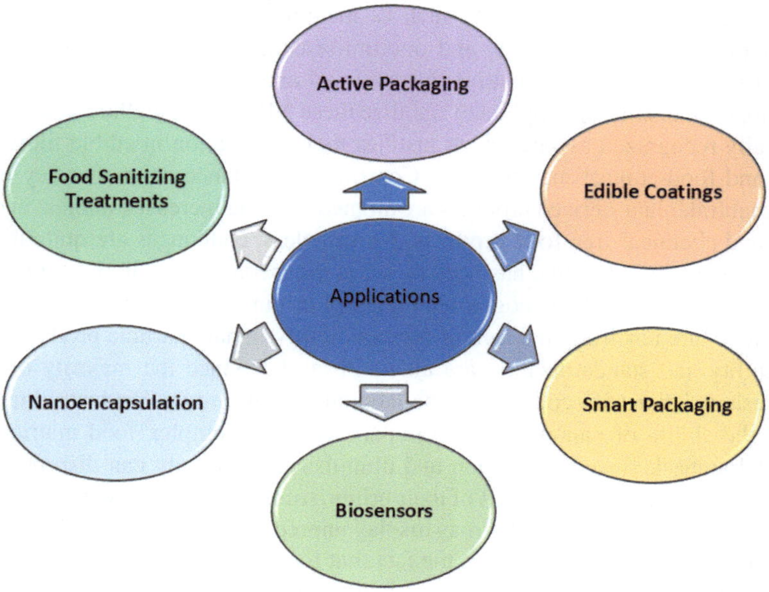

Fig. 4.1 Applications of underutilized plant-derived nanoparticles in food safety

controlled release of bioactive plant compounds with the use of starch- or cellulose-based polymeric nanoparticles are enhanced by nanoencapsulation methods, increasing their functional use in food systems. Moreover, there are also intelligent packaging or biosensor sensors with carbon dots and AgNP-based sensors that can detect changes in microbial contamination or toxins on food in real-time, providing real-time non-destructive insight into food quality. These applications together carry precise evidence of the opportunities possessed by nanomaterials of plant-related origin in ensuring security and quality control throughout the food supply chain [27, 31]. Applications of underutilized plant-derived nanoparticles in food safety are shown in Fig. 4.1.

4.5 Advantages and Challenges of Underutilized Plant-Derived Nanoparticles

The use of the underutilized plant-based nanoparticles (NPs) in food safety offers a distinct combination of sustainability, efficacy, and consumer acceptance. However, their large-scale use is limited by challenges such as non-standardized synthesis, safety concerns, scalability issues, and limited regulatory approval. On the good side of the matter, these NPs are greener and sustainable, as they correspond to the principles of the circular economy as they give much more value to the by-products of agriculture, e.g., the fruit peels, seeds, and pomace [13]. They possess multi-target antimicrobial effects such as membrane disruption, oxidative stress production, and DNA damage, and quorum-sensing, assessing which lowers the risk of bacterial resistance relative to traditional antibiotics [2]. Moreover, numerous biopolymers that are applied to stabilize these NPs (pectin, cellulose, starch) are generally recognized as safe which justifies their application in edible films, coatings, and food-contact materials [9]. Consumer acceptance is another key benefit since natural, plant-derived products are aligned with the increased interest in clean-label and chemical-free food products. Nevertheless, challenges are quite distinct. One of the main issues that has been raised is the inconsistency in the composition of plant extracts based on geographic region of origin, time of year, and the extraction procedure resulting in a lack of reproducibility in nanoparticle production [8]. Scalability and standardization are also still hurdles since the majority of green synthesis reactions are confined to the laboratory with minimal industrial transfer. Also, the ability of nanoparticles to remain stable in complex food matrices is a major drawback since aggregation and diminished bioactivity can disrupt antimicrobial functionality. Regardless of their origin, safety and toxicological assessment is necessary since nanoparticles may display unpredictable bio-distribution and toxicity in the nanoscale [10]. Lastly, the fact that food nanotechnology has no regulatory frameworks and labeling guidelines is a significant bottleneck to commercialization [32]. Comprehensively, even though underutilized plant-derived nanoparticles have remarkable potential as a sustainable antimicrobial agent in food systems, addressing these issues with sophisticated characterization, standardization protocols, and regulatory clarity will be a major step in ensuring the transfer of laboratory success to industrial practice [33].

4.6 Future Perspectives

- Although conventional metal and metal-oxide nanoparticles exhibit efficient antimicrobial and antibiofilm effects, they are associated with limitations, namely, cytotoxicity, low biodegradability, high cost of production, improper stability at different conditions of pH/ionics, and possible adverse interactions with food components.
- Regulatory issues and generalization of oxidative stress only add to the limitations of using them in food systems, which heightens a motivation to seek safer biogenic, plant-derived nanoparticle solutions.
- A significant innovation potential can be seen in underutilized plant-derived nanoparticles, especially in terms of using them in hybrid materials (polymer-metal composite or nano-biohybrids), which enhance plant-derived nanoparticles' stability and antimicrobial efficacy in food matrices [2].
- Plant-based nanotechnology, together with other preservation methods (hurdle technology) like cold storage, modified atmosphere packaging, and natural antimicrobials, could help to have greater control over the microbials and increase shelf life.

- Omics-based methods (genomics, metabolomics, proteomics) facilitate the systematic identification of novel plant-based phytochemicals that are reducing, stabilizing, or functionalizing agents in nanoparticle assembly to increase the library of bioactive compounds [33].
- Safety testing is also necessary since plant-based NPs can act in a different way at the nanoscale; the safety regulations and toxicological analysis (FAO/WHO 2022) will be critical to the study and adoption by the industry and consumers.
- Their utilization in the future is expected to be in the form of nanocarriers that can be used in the delivery of bioactives, probiotics, or nutraceuticals with the object of developing personalized and functional foods into the global food chain [34].
- Nanomaterials that are stimuli-responsive allow adhesive antimicrobials to be released in response to temperature, moisture, microbial metabolites, or pH variations during spoilage to enhance the safety and minimize unnecessary accumulation of chemicals.
- It can be integrated in smart packaging and electronic tracking, including carbon-dot nanosensors on fruit waste, to detect real-time presence of pathogens or spoilage markers, which will improve the final outcome of transparency and consumer trust [24].
- Scalable, climate-resilient, and environmentally responsible development of plant-based nanotechnology in food preservation will be the efforts of machine learning, sustainable production models built on agri-waste, biodegradable biopolymer incorporation, and cooperation between scientists, regulators, and the industry [31, 35].

4.7 Conclusion

Plants that are under-utilized are also a viable and inexpensive source of nanoparticles, which in turn is eco-friendly as compared to the traditional methods of producing nanoparticles utilizing chemicals and physical forces. The abundance of phytochemicals in these plants is such that nanoparticles derived have inherent antibacterial properties which are very applicable in the treatment of food-borne infections like *Salmonella*, *E. coli*, and *Listeria*. Case studies prove that plant-derived NPs would be successfully implemented in food packaging, edible coating, sanitizing agent, and nanoencapsulation systems to minimize spoilage and microbial contamination. Their multi-target anti-bacterial activities, such as disrupting the membrane, producing ROS, and preventing biofilm formation, demonstrate their advantages over single-target chemical preservatives. Nonetheless, variability of plant extracts, regions without stability in nanoparticles, ability to scale, and regulatory acceptance remain an issue of concern. To open these gaps to maximize their potential, interdisciplinary studies are required to implement the ideas of advanced nanotechnology, safety factors, and clear laws. Lastly, untapped nanoparticles on the plant-based are a disruptive potential in food safety and preservation, a convergence of green chemistry, sustainability, and consumer-seeking clean-label. They could play a leading role in the new generation of antimicrobial food technologies that have further research and regulatory support. In addition to their direct antimicrobial properties, synthesized nanoparticles based on plants that are not used extensively lead to a more significant change in the way food safety issues are handled. The plant-based nanostructures provide a simultaneous approach in which the antimicrobial response, environmental impact, and resource optimization are met. As compared to the traditional chemical-treated antimicrobial agents, which are mostly dependent on synthetic chemicals or energy-intensive reactions, the green-synthesized nanoparticles involve the utilization of naturally found phytochemicals, which negatively impact the environment and decrease the cost of production. This makes them a viable approach to industries that intend to achieve the sustainability goals without interfering with microbial control.

Author's Contribution RK—Writing original draft, data collection; RK—Data interpretation, conceptualization; JA—Final formatting, final approval; MQ—Data collection.

Ethical Approval The author declares no competing interests.

Conflict of Interest Not applicable.

References

1. WHO (2022) WHO global strategy for food safety 2022–2030: towards stronger food safety systems and global cooperation. WHO. https://iris.who.int/handle/10665/363475
2. Bhardwaj SK, Deep A, Bhardwaj N et al (2023) Recent advancements in nanomaterial-based optical detection of food additives: a review. Analyst 148(21):5322–5339. https://doi.org/10.1039/D3AN01317K
3. Poonia A, Dhewa T (2022) Edible food packaging. Springer Singapore. https://doi.org/10.1007/978-981-16-2383-7
4. Jannat K, Paul AK et al (2021) Nanotechnology applications of flavonoids for viral diseases. Pharmaceutics 11:1895. https://doi.org/10.3390/pharmaceutics11121895
5. Khalil AM, Saied S et al (2024) Green biosynthesis of bimetallic selenium–gold nanoparticles using *Pluchea indica* leaves and their biological applications. Front Bioeng Biotechnol 11:1294170. https://doi.org/10.3389/fbioe.2023.1294170
6. Rama P, Mariselvi T, Sundaram R, Muthu K (2023) Eco-friendly green synthesis of AgNP from *Aegle marmelos* leaf extract and their antimicrobial, antioxidant, anticancer and photocatalytic degradation activity. Heliyon 9(6):e16277. https://doi.org/10.1016/j.heliyon.2023.e16277
7. de Sousa MS, Schlogl AE et al (2023) Nanotechnology in packaging for food industry: past, present, and future. Coatings 13(8):1411. https://doi.org/10.3390/coatings13081411
8. Awlqadr FH, Majeed et al (2025) Nanotechnology-based herbal medicine: preparation, synthesis, and applications in food and medicine. Agric Food Res 19:101661. https://doi.org/10.1016/j.jafr.2025.101661
9. Wahab S, Khan T et al (2021) Mechanistic aspects of plant-based AgNP against multi-drug-resistant bacteria. Heliyon 7(7):e07564. https://doi.org/10.1016/j.heliyon.2021.e07448
10. Singh P, Pandit S et al (2021) AgNP produced from *Cedecea* sp. exhibit antibiofilm activity and remarkable stability. Sci Rep 11:12619. https://doi.org/10.1038/s41598-021-92006-4

11. Ashraf MA, Khan AS et al (2025) Innovative technologies in postharvest management of fruits and vegetables: a review. Eur Food Res Technol 1:19. https://doi.org/10.1007/s00217-025-04861-4
12. Rai M, Feitosa CM et al (2025) Harnessing bioactive nanocurcumin and curcumin nanocomposites to combat microbial pathogens: a comprehensive review. Crit Rev Biotechnol 1:23. https://doi.org/10.1080/07388551.2025.2458006
13. Luzala MM, Muanga J et al (2022) A critical review of the antimicrobial and antibiofilm activities of green-synthesized plant-based metallic nanoparticles. Nanomaterials 12(11):1841. https://doi.org/10.3390/nano12111841
14. Ullah F, Ayaz M, Sadiq A et al (2020) Potential role of plant extracts and phytochemicals against foodborne pathogens. Appl Sci 10(13):4597. https://doi.org/10.3390/app10134597
15. Elshaer S, Shaaban MI (2023) Antibiofilm activity of biosynthesized silver and copper nanoparticles using Streptomyces S29. AMB Express 13:139. https://doi.org/10.1186/s13568-023-01647-3
16. Wood JE, Gill BD (2025) Introduction to mass spectrometry for metabolite profiling. In: Bioanalytical techniques: principles and applications, vol 28, issue 3. Wiley, pp 340–357. https://doi.org/10.1002/9781394314133.ch12
17. Singh NB, Chaudhary RG et al (2023) Green synthesized nanomaterials for safe technology in sustainable agriculture. Curr Pharm Biotechnol 24(1):61–85. https://doi.org/10.2174/1389201024666230206154924
18. Dakal TC, Kumar A et al (2016) Mechanistic basis of antimicrobial actions of silver nanoparticles. Front Microbiol 7:1831. https://doi.org/10.3389/fmicb.2016.01831
19. Havelaar AH, Kirk MD, Torgerson PR, World Health Organization Foodborne Disease Burden Epidemiology Reference Group et al (2015) World Health Organization global estimates and regional comparisons of the burden of foodborne disease in 2010. PLoS Med 12(12):e1001923. https://doi.org/10.1371/journal.pmed.1001923
20. Siddiqi KS, Husen A et al (2018) A review on biosynthesis of AgNP and their biocidal properties. NanoBiotechnology 16(1):14. https://doi.org/10.1186/s12951-018-0334-5
21. Singh A, Singh NÁ et al (2018) Zinc-oxide nanoparticles: a review of their biological synthesis, antimicrobial activity, uptake, translocation and biotransformation in plants. Mater Sci 53(1):185–201. https://doi.org/10.1007/s10853-017-1544-1
22. Chandraker SK, Kumar R (2024) Biogenic biocompatible silver nanoparticles: a promising antibacterial agent. Biotechnol Genet Eng 40(4):3113–3147. https://doi.org/10.1080/02648725.2022.2106084
23. Ahmed S, Ahmad M et al (2016) A review on plant-extract mediated synthesis of AgNP for antimicrobial applications: a green expertise. Adv Res 7(1):17–28. https://doi.org/10.1016/j.jare.2015.02.007
24. Sharma R, Lata P et al (2024) Valorisation of agricultural waste and their role in green synthesis of value-added nanoparticles. Environ Technol Rev 13(1):40–59. https://doi.org/10.1080/21622515.2023.2283412
25. Tailor G, Chaudhary J, Mehta R et al (2024) Recent methods for biogenic synthesis of metal nanoparticles and their applications. In: Recent advancements in multidimensional applications of nanotechnology. Bentham Science, pp 203–224. https://doi.org/10.2174/9789815238846124010l
26. Kaiser KG, Delattre C et al (2023) Nanosilver: an old antibacterial agent with great promise in the fight against antibiotic resistance. Antibiotics 12(8):1264. https://doi.org/10.3390/antibiotics12081264
27. Verma D, Yadav AK et al (2022) Prospects of nanomaterial-based biosensors: a smart approach for bisphenol-A detection in dental sealants. J Electrochem Soc 169(2):027516. https://doi.org/10.1149/1945-7111/ac4761
28. Vanlalveni C, Lallianrawna S, Biswas A et al (2021) Green synthesis of silver nanoparticles using plant extracts and their antimicrobial activities: a review of recent literature. RSC Adv 11(5):2804–2837. https://doi.org/10.1039/D0RA09941D

29. Tiwari S, Dubey NK (2022) Nanoencapsulated essential oils as novel green preservatives against fungal and mycotoxin contamination of food commodities. Curr Opin Food Sci 45:100831. https://doi.org/10.1016/j.cofs.2022.100831
30. Hajipour MJ, Fromm KM et al (2012) Antibacterial properties of nanoparticles. Trends Biotechnol 30(10):499–511. https://doi.org/10.1016/j.tibtech.2012.06.004
31. Dhatwalia J, Kumari A et al (2022) *Rubus ellipticus* Sm. fruit extract mediated zinc-oxide nanoparticles: a green approach for dye degradation and biomedical applications. Materials 15(10):3470. https://doi.org/10.3390/ma15103470
32. World Health Organization (2024) Pesticide residues in food 2022. Joint FAO/WHO meeting on pesticide residues. Evaluation Part II–Toxicological. World Health Organization
33. Baranwal J, Barse B, Fais S et al (2022) Biopolymer: a sustainable material for food and medical applications. Polymers 14(5):983. https://doi.org/10.3390/polym14050983
34. Iravani S (2011) Green synthesis of metal nanoparticles using plants. Green Chem 13(10):2638–2650. https://doi.org/10.1039/c1gc15386b
35. Ismail T, Sestili P, Akhtar S (2012) Pomegranate peel and fruit extracts: a review of potential anti-inflammatory and anti-infective effects. J Ethnopharmacol 143(2):397–405. https://doi.org/10.1016/j.jep.2012.06.018

Chapter 5
Underutilized Plant-Derived Nanoparticles Against Water-Borne Bacteria

Sakshi Sharma and Ritesh Banerjee

Abstract Waterborne bacterial disorders pose a serious threat to global health, particularly in regions with limited access to safe drinking water. The increase in antibiotic resistance has intensified the need for innovative antimicrobial strategies. Among these, green-synthesized nanoparticles, particularly those prepared from plant sources, provide a sustainable and effective solution. This chapter focuses on the green synthesis of nanoparticles using plants, emphasizing their biocompatibility, environmental-friendliness, and low-cost production. Detailed antibacterial mechanisms, like reactive oxygen species generation, cell membrane disruption, and ion leaching, are addressed, including their varying efficacy against Gram-positive and Gram-negative bacteria. Specific focus is given to the application of silver, zinc oxide, copper, and gold nanoparticles against key waterborne pathogens including *Escherichia coli*, *Vibrio cholerae*, and *Salmonella typhi*. The chapter also assesses the incorporation of these nanoparticles into water treatment systems, while addressing critical concerns like nanotoxicity, environmental persistence, and biosafety. By integrating nanotechnology and natural plant compounds, this emerging field holds potential for sustainable solutions against waterborne bacterial problems.

Keywords Sustainable solutions · Green synthesized nanoparticles · Water-borne bacteria

S. Sharma · R. Banerjee (✉)
School of Biological and Environmental Sciences, Shoolini University,
Solan, Himachal Pradesh, India
e-mail: riteshbanerjee@shooliniuniversity.com

© The Author(s), under exclusive license to Springer Nature Switzerland AG 2026
E. Nepovimova et al. (eds.), *Underutilized Plants Based-Nanoparticles: Herbal Remedy to Treat Human Diseases*, https://doi.org/10.1007/978-3-032-20384-7_5

5.1 Introduction

Water is essential to life, and making sure everyone has access to clean and safe drinking water is a critical global need. The World Health Organization (WHO) reports that more than five million humans die each year from water-related diseases, with over 50% of these deaths resulting from microbial intestinal infections, most notably cholera [1]. Many different types of pathogenic microorganisms, such as bacteria, viruses, and protozoa, can be found in contaminated surface waters. There have been many outbreaks of *Escherichia coli* O157:H7, the most severe of which happened in Walkerton, Ontario, Canada in the spring of 2000 and led to over 2300 illnesses and six fatalities. The quantity of outbreaks documented globally indicates that drinking water contamination is still a major source of disease [2].

In the twenty-first century, nanotechnology has surfaced as a groundbreaking scientific field offering creative solutions in a variety of fields, including the environment, agriculture, food, biotechnology, medications, etc. [3]. The term "nanoparticles" (NPs) refers to particles that are 100 nm or smaller in at least one dimension. It is crucial to note that, in contrast to the similar bulk material, NPs exhibit distinct and frequently better physico-chemical characteristics. The size of NPs determines their physical and chemical characteristics, whereas bulk materials typically have consistent physical characteristics independent of their size [4, 5]. Physical and chemical procedures are typically used in conventional synthesis techniques, which have drawbacks in terms of energy consumption and the release of hazardous compounds. The utilization of biological processes to prepare NPs is proven to be a viable and sustainable substitute. Their benefits over conventional techniques include lower energy use, affordability, and environmental friendliness [6].

Microbiological examination of the water resources is a crucial part of water quality and human health due to the rate and extent of urbanization, which has increased the utilization of surface water for human consumption and recreation [7]. Nowadays, numerous techniques, like ozonization, UV treatment, and chlorination, are used to disinfect water. However, when used on a large scale, these techniques have disadvantages such as excessive energy consumption and the production of hazardous disinfection byproducts [8]. Nanotechnology-based solutions are now a practical way to deal with water quality and quantity problems in both developed and developing countries [9]. The use of NPs in water treatment systems can greatly increase disinfection effectiveness by improving some of the drawbacks of conventional techniques, like the high energy requirements of UV radiation or the production of hazardous byproducts during chlorination. NPs use a range of complex methods that are strongly influenced by physicochemical characteristics of the specific nanomaterial to resist microbial contamination [10]. As a result, NPs have attracted a lot of interest lately because of their exceptional capacity to eradicate waterborne bacteria, providing a potential alternative to conventional methods [11]. Despite their numerous advantages in treating waterborne bacteria and improving the quality of water, NPs are still not being synthesized from underutilized plants. An excellent illustration of this is that 10% of the world's plant species are found in

South Africa, where more than 3000 plants have been shown to have amazing antibacterial properties. In spite of this, many plants are still underutilized for the synthesis of NPs [12]. In the current chapter, we have mentioned the synthesis methods of NPs along with the mechanisms by which NPs work against various water-borne bacteria. We have also highlighted the nanotoxicity, environmental persistence, and biosafety.

5.2 Synthesis of Nanoparticles

Organic NPs are the solid particles made from organic compounds such as lipids or polymers with a diameter ranging from 10 nm to 1 μm. A few widely known organic NPs are dendrimers, liposomes, micelles, ferritin, etc. Organic NPs are eco-friendly, biodegradable, non-toxic, cost-effective, and better suited for biomedical fields. Whereas, inorganic NPs are not made of carbon. It consists of metal and metal oxides [13]. Nanomaterials are prepared by two methods, the bottom-up method and the top-down method as shown in Fig. 5.1. The bottom-up, or self-assembly, approach of nanofabrication involves the utilization of chemical or physical forces at the nanoscale to combine building blocks into functional structures. In the bottom-up method, particles are made from the atomic to cluster to the NPs level [14]. Chemical and biological methods are known as a bottom-up approach. In top-down methods, bulk materials are fragmented to create nano-structured materials. They are also called physical methods [3].

Nowadays, more attention is given to biological methods, which include the utilization of natural reducing, capping, and stabilizing agents. It doesn't use toxic,

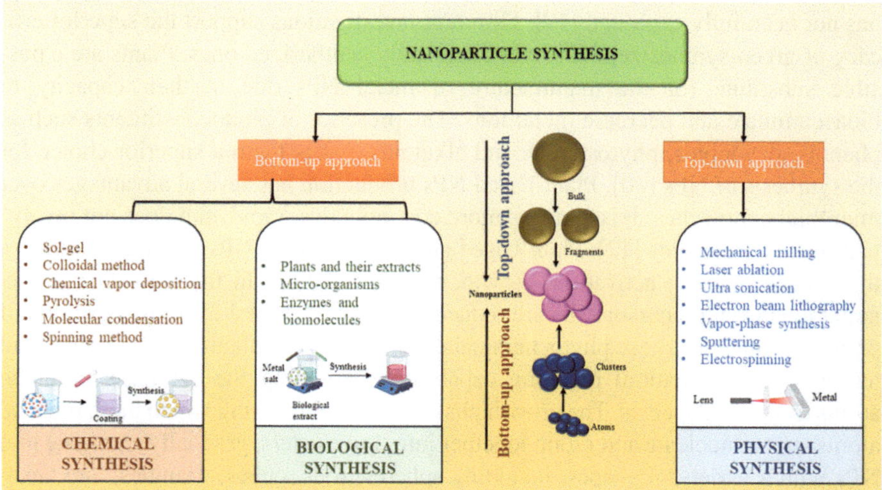

Fig. 5.1 Synthesis of nanoparticles

expensive chemicals and high-energy utilization machinery [15]. Although conventional methods have been in use for many years, scientists have demonstrated that green approaches are more effective in producing NPs and have the advantages of being easy to characterize, inexpensive, and less likely to fail. Numerous environmental stresses have been demonstrated by the physical and chemical methods of manufacturing NPs because of their hazardous metabolites [16]. The reducing potential of the chemicals found in living things is used in the green synthesis of NPs. Microorganisms and plants are accessible, global, and simple to work with. Furthermore, the manufacturing of NPs can be shifted to the tenets of a circular economy by using the biomass waste in synthesizing green NPs [6]. The biosynthetic method is a safe, biocompatible, environmentally safe green technique to prepare NPs with plants and microorganisms for different applications. Another advantage of plant-assisted NPs synthesis is that the kinetics for this route is much higher than other biosynthetic techniques, equivalent to chemical NPs preparation. Different plant parts like fruit, leaf, stem, and root have been readily utilized for green synthesis of NPs due to their extraordinary phytoconstituents [17].

Following the preparation of NPs, it is essential to examine their morphology and other subtle structural aspects using a variety of spectroscopic techniques. The most common systems are: UV–vis absorption spectroscopy, X-ray diffraction, Fourier transmission infrared spectroscopy, dynamic light scattering, energy dispersive X-ray examination, scanning electron microscopy, and transmission electron microscopy [16, 18].

5.2.1 Underutilized Plant-Mediated Synthesis of Nanoparticles

The word underutilized plant is commonly used for plant species whose potential has not been fully exploited [19]. Different investigations support the superior efficacy of green-synthesized NPs over chemically synthesized ones. Plants are a possible substitute for the manufacture of metal NPs due to their capacity to bioaccumulate and decrease metal ions. The presence of phytoconstituents such as phenols, flavonoids, phytosterols, and alkaloids makes them a superior choice for the synthesis of NPs [20]. Plant-based NPs production has several advantages over microbial approaches: it is simpler, more cost-effective, rapid, and does not involve pathogenic organisms [12]. Plant-based green synthesis of NPs generally follows a three-stage process: activation, growth, and termination. In the activation phase, metal ions from precursor salts are reduced to their elemental form by the action of phytoconstituents. These phytochemicals, like flavonoids, alkaloids, phenolics, and terpenoids, have natural reducing capabilities, initiating the synthesis of metal atoms from ionic species. The growth phase comes next, during which these reduced atoms start to nucleate and group together into tiny clusters, gradually changing into NPs with a variety of shapes, including spheres, rods, cubes, triangles, and hexagons. During this phase, ongoing biological reduction may continue, and the developing NPs gain thermodynamic stability. The size and shape of NPs can be affected by reaction parameters such as temperature, pH, concentration of plant extract, and metal salt. Finally, in the termination stage, the NPs get a stable configuration and are capped by plant metabolites, which stop agglomeration and preserve their structural integrity. This capping also contributes to the biological activity and surface functionality of the NPs, making them ideal for being used in microbial disinfection and water purification [21, 22]. NPs from underutilized plants have performed better than their bulk and chemically generated counterparts. For instance, zinc oxide NPs (ZnO NPs) made from Aloe vera extract have demonstrated greater antibacterial activity. Similarly, ethnobotanically utilized plants from South Africa such as cassava, *Garcinia kola*, *Eleusine coracana*, and *Lablab purpureus* are increasingly being recognized for their potential in NPs synthesis due to their unique phytochemical properties [23, 24]. Silver NPs (AgNPs) produced from plants have also demonstrated intriguing antibacterial properties, suggesting that they may find value in a range of biomedical applications. The phytochemicals found in plant extract determine the type of NPs that can be produced from plants [25]. There are many medicinal plants with amazing antibacterial activities because of the myriad of phytochemicals present in them. For example, bamboo, along with being a superfood, has multiple phytochemicals like phenols, flavonoids, phytosterols, alkaloids, and polysaccharides that may provide desirable health benefits. The majority of plants have amazing antioxidant, antibacterial, antibiotic, and disinfectant activities [20]. Another example of an underutilized plant is *Kaempferia angustifolia* that is native to Southeast Asia. The NPs made from the rhizome of this plant have been utilized in an antibacterial study, and according to the results, it can be explored in pharmaceutical industries for treating multidrug-resistant human pathogenic bacteria [26]. A few more examples of underutilized plants used for NPs synthesis with antibacterial activities are discussed in Table 5.1.

5.3 Antibacterial Mechanism of Nanoparticles Against Waterborne Pathogens

Most bacteria, with the exception of a few species, have an exterior protective cell wall and membrane. They can be categorized as either "Gram-positive" or "Gram-negative" based on the makeup of their cell walls. The homogeneous and comparatively thick (20–80 nm) wall of Gram-positive bacteria is made up of several peptidoglycan layers that are interspersed with teichoic and teichuronic acids, which are polymers of alcohols and phosphates. They only permit the entry of hydrophilic molecules due to their highly polar nature. While Gram-negative bacteria have an extremely thin (less than 10 nm) and complicated wall, with an outer membrane that contains an exterior lipopolysaccharide and an internal phospholipid layer encircling the peptidoglycan. The outer membrane makes the wall extremely selective and prevents water and other tiny hydrophilic molecules from passing through [44, 45]. Gram-positive and Gram-negative bacteria react differently to metal NPs and

Table 5.1 List of NPs synthesized from underutilized plants with antibacterial activity

Plant name	Part used	NPs synthesized	References
Baccaurea ramiflora	Fruit	Ag NPs	[27]
Syzygium cumini	Seeds	Ag NPs	[28]
Detarium microcarpum	Leaves	Ag NPs	[29]
Leonotis ocymifolia	Leaves	ZnO NPs	[23]
Elaeagnus pyriformis	Fruit	Ag NPs	[27]
Bridelia ferruginea	Stem bark	ZnO NPs	[30]
Crescentia cujete	Leaves	CuONPs	[31]
Amaranthus cruentus	Shoot	Ag NPs	[32]
Borreria Verticillata	Leaves	Cu NPs, Fe NPs, Zn NPs	[33]
Salvia hispanica	Seeds	ZnO NPs, Ag NPs	[34, 35]
Eleusine coracana	Seeds	CaO NPs	[36]
Eragrostis tef	Seeds	Ag NPs	[37]
Vitellaria paradoxa	Seeds	Ag NPs	[37]
Dovyalis caffra	Leaves, fruits	CuO NPs, ZnO NPs, Ag NPs	[38, 39]
Terminalia mantaly	Leaves, stem, bark	Ag NPs, ZnO NPs	[40, 41]
Cassia fistula	Leaves	ZnO NPs	[42]
Detarium microcarpum	Leaves	Ag NPs	[43]

Table 5.2 List of a few waterborne bacterial pathogens

Bacteria	Diseases	References
Campylobacter jejuni, C. coli	Gastroenteritis	[47]
Legionella pneumophila	Pneumonia	[7]
Vibrio cholerae	Cholera	[47]
Helicobacter pylori	Gastritis, peptic and duodenal ulcer disease, and gastric cancer	[48]
Escherichia coli	Diarrhea	[49]
Mycobacteria spp.	Pulmonary and cutaneous disease, disseminated infections	[7]
Shigella spp.	Shigellosis	[47]
Klebsiella pneumoniae	Pneumonia	[50]
Pseudomonas aeruginosa	Nosocomial pneumonia	[51]
Salmonella Typhi	Typhoid fever	[47]
Staphylococcus aureus	Skin infection	[52]

other antimicrobial agents because of their distinct structural arrangements. Several investigations have shown that due to their distinct organization and structure, which allow released ions to enter more easily, Gram-positive bacteria are more resistant to NPs than Gram-negative bacteria [46]. A few examples of waterborne bacteria are mentioned in Table 5.2.

Metal-based NPs act as non-specific antibacterial agents as they don't bind to a specific receptor in the bacterial cell. Metal-based NPs restrict the development of resistance by bacteria and increase the range of antibacterial activity [53, 54]. NPs' exposure to cells of bacteria can result in membrane damage from NP adsorption, which is occasionally followed by cell penetration. Cell wall depolarization brought on by NP adsorption alters the normally negative charge of the wall, making it more permeable. Using a laser scanning confocal microscope, it has been reported that the bacterial cell wall becomes fuzzy, signifying cell wall disintegration due to the NPs [55]. Gold (Au) and Ag NPs at low concentrations are less toxic to normal cells, but more harmful to bacteria. Due to this, Ag and Au NPs have been mostly studied as antimicrobial agents [56]. Environmental factors, such as pH and NP dissolving rate, have been shown in numerous studies to have a substantial impact on the antibacterial property of NPs. Saliani et al. demonstrated that when pH dropped from 7 to an acidic pH, ZnONPs' ability to suppress bacterial growth increased [57]. The antibacterial mechanisms of NPs and the work done against water-borne bacteria are further discussed in detail.

5.3.1 Mechanistic Insights and Studies on Silver Nanoparticles (AgNPs)

Several studies have proposed a bimodal mechanism underlying the antibacterial activity of AgNPs. Initially, AgNPs disrupt the integrity of the bacterial cell wall, enabling their subsequent internalization. After entering the cell, they cause ROS to be generated, which in turn interferes with vital cellular processes like ATP synthesis and DNA replication. The ROS not only overpowers the bacterial antioxidant defense system but also contributes to early-stage membrane damage [58]. The sustained release of silver ions from AgNPs further increases their antibacterial efficacy, ultimately causing bacterial cell death. These NPs readily attach to the phosphorus and sulfur-rich proteins present on the cell membrane of bacteria, facilitating their entry into the cytoplasm. Within the cell, AgNPs interact with vital biomolecules like DNA and ribosomal proteins, thus inhibiting protein synthesis. This interference causes ribosomal disassembly, enzyme inactivation, and impairment of the respiratory chain and cell division machinery, hence resulting in cell death [59]. The mechanism is displayed in Fig. 5.2. In addition to their physical interactions, the antibacterial action of NPs is frequently mediated by the leaching of metal ions like Ag^+, which enter bacterial cells and interfere with essential biomolecules like enzymes and nucleic acids [46]. In a study, AgNPs were rapidly synthesized from the leaf extract of *Acalypha indica.* Their antibacterial activity showed effective inhibitory property against water-borne bacteria, *E. coli* and *Vibrio cholerae*. The biosynthesized AgNPs demonstrated potent antimicrobial activity, as evidenced by minimum inhibitory concentration (MIC) assessments, alterations in membrane permeability, and disruption of respiratory activity in treated bacterial cells [60]. The antibacterial activity of the chemically prepared AgNPs was studied using the agar well diffusion method against *Enterococcus* spp. and *E. coli*. The results

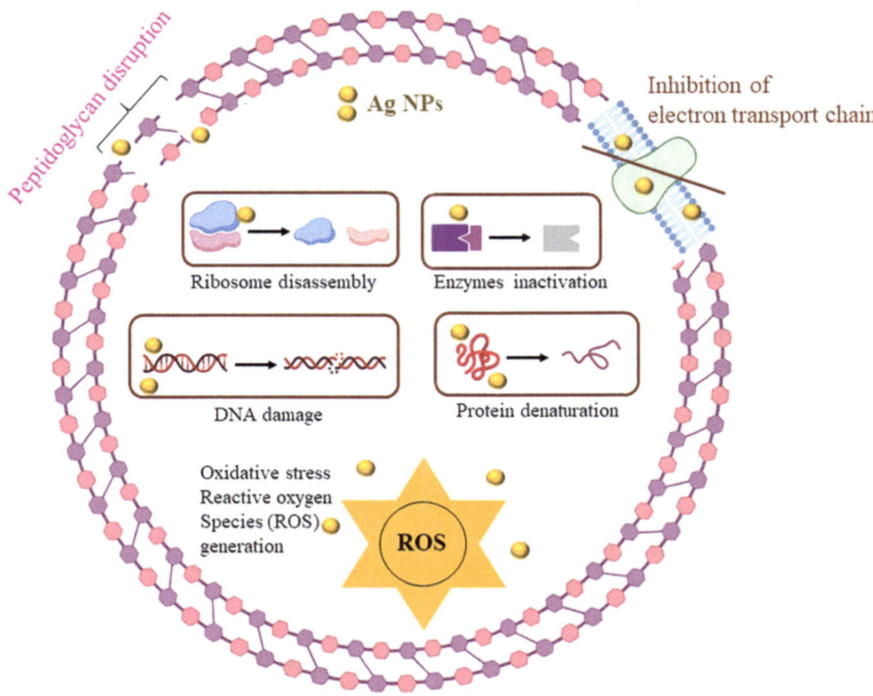

Fig. 5.2 Antibacterial mechanism of silver nanoparticles (Ag NPs)

highlighted that AgNPs can be utilized as effective growth inhibitors against waterborne bacterial pathogens, making them a potential candidate for different water filters and antimicrobial applications [61]. In another study, the antimicrobial potential of AgNPs prepared from the leaves of *S. cumini* was studied against *S. aureus.* It has shown dose-dependent antimicrobial and antibiofilm activities against the bacteria, with minimal cytotoxicity to oral keratinocytes [62].

5.3.2 Mechanistic Insights and Studies on Zinc Oxide Nanoparticles (ZnONPs)

The most prevalent and generally acknowledged mechanism for ZnONPs' antibacterial action is ROS. In order to provide the bactericidal action, ROS can break the chemical interactions that hold bacteria's organic substances together. Among these, the negatively charged peroxide is unable to penetrate the cell membrane, OH^- aggregates on the surface of bacterial cell membranes and destroys them, and H_2O_2 can enter the cell membrane and damage it, destroying the membrane's DNA and proteins, which has a bactericidal effect [63]. In aqueous solution, ZnO can release Zn^{2+} gradually. Zn^{2+} has the capability to pass by cell membranes, denaturing proteins, and preventing cell division. Furthermore, cellular respiratory disorders can result from Zn^{2+}'s destruction of the electron transport system. Due to Zn^{2+}'s electrostatic attraction to the negatively charged bacterial cell membrane surface, the charge balance on the membrane will be disrupted, causing significant cell deformation and ultimately bacterial lysis [64].

5.3.3 Mechanistic Insights and Studies on Copper Nanoparticles (CuNPs)

The antibacterial action of copper-based NPs (CuNPs) primarily involves the production of ROS, which play an essential part in inducing oxidative stress, leading to destruction of bacterial cell walls and membranes. Additionally, these NPs disturb vital biological functions by interacting with intracellular proteins and DNA. CuNPs are adsorbed onto the bacterial surface by electrostatic interactions, forming strong bonds that disrupt cell wall integrity. This disruption increases membrane permeability, facilitating the internalization of NPs and amplifying their bactericidal effects [65]. A recent study demonstrated the effectiveness of CuNPs synthesized from the pod of water caltrop against *E. coli* and *S. aureus*, using the well diffusion technique to highlight their efficacy for sustainable antibacterial applications [66].

5.3.4 Mechanistic Insights and Studies on Gold Nanoparticles

The antibacterial activity of AuNPs extends beyond ROS generation. Studies have shown that AuNPs can interrupt bacterial energy metabolism and transcriptional processes. Their broad-spectrum antimicrobial mechanism includes the induction of apoptosis, DNA damage, disruption of the cell membrane, impairment of the electron transport chain, ROS production, and interference with key metabolic pathways, notably ATP synthesis [67]. A recent study has shown that AuNPs were prepared from the leaves of *S. cumini.* The prepared NPs have shown excellent antibacterial activities against multidrug-resistant waterborne bacteria, including *E. coli* and *S. aureus*, showing their potential against waterborne bacteria.

5.4 Toxicological Considerations and Environmental Risks

Nanotechnology has an impact on the environment, just like any other technology, and this impact must take into account related to issues with environmental health, human toxicity, and embodied energy. Although a great deal of study has been done on safe nanomaterial compositions, more has to be done to comprehend the links between nanomaterial structure, attributes, and hazards because these relationships can affect exposure pathways and toxicity [9]. The risk associated with manufactured NPs is primarily determined by two interrelated factors: toxicity and exposure. The toxicity of nanomaterials is impacted by their surface chemistry and chemical stability, which control how these NPs interact with biological systems. Equally significant is the potential for exposure, including the transfer, bioavailability, and accumulation of NPs in the environment, particularly by wastewater pathways. One of the major environmental concerns associated with emerging nanotechnologies is the potential bioaccumulation of NPs in natural ecosystems. Interestingly, the cytotoxic effects of certain NPs can be reduced through the application of organic surface coatings, which help modify interactions with cells and decrease adverse effects [68]. Natural organic matter can have a substantial impact on the toxicity of metallic NPs by changing various physicochemical properties. These include increasing suspension stability, modifying the bioavailability of metal ions, and modulating electrostatic and steric interactions between NPs and biological systems. Furthermore, natural organic matter can affect the ROS generation induced by metallic NPs, which is a major factor in their cytotoxicity. In aquatic environments, NPs' toxicity often manifests through adsorption onto cell surfaces and subsequent disruption of membrane transport processes. Given these complex interactions, it is important that all engineered NPs undergo comprehensive toxicological assessments prior to their application in human health or the environment [69]. Preclinical investigations are conducted with the possible test material in the first phase. These studies are often conducted in vitro, with cell lines originating from target and off-target tissues, and subsequently in vivo, using living higher organisms. Multiple investigations using numerous NPs have shown the significance of factors such as the particle's overall size, shape, aggregation state, chemical composition, biostability, and surface chemistry, among others. Thus, from a researcher's perspective, a thorough description of these NPs' physico-chemical characteristics, which form the foundation of their physical identity, is essential for determining the toxicity that results [70].

5.5 Challenges and Future Perspective

The toxicity of plant-mediated NPs has been assessed using a variety of biological models, including plants, fish, mammals, microbes, and mammalian cells. However, none of these models has been able to adequately pinpoint the precise mechanism underlying the toxicity of these NPs. Therefore, it is advised that significant efforts be made to evaluate the safety and effectiveness of these formulations of natural metallic NPs in clinical research [12]. The presence of metallic NPs in wastewater and sludge from ineffective industrial waste management has increased ecological significance in recent decades, in addition to the environmental impact of NPs from their utilization in agriculture. Consequently, it has been determined that the existence of NPs in industrial waste may contribute to the progression of antibiotic resistance in the environment. The creation of such resistance, which encompasses the spread of bacteria and genes resistant to antibiotics, is, nevertheless, poorly understood. Therefore, understanding how nanomaterials contribute to the emergence of antibiotic resistance in environmental systems may be the main goal of future research [71].

5.6 Conclusion

The rising prevalence of waterborne diseases caused by bacteria, exacerbated by urbanization and inefficient disinfection systems, underscores the urgent need for novel and sustainable antimicrobial strategies. Nanotechnology offers promising answers, and among various types of NPs, plant-derived NPs have emerged as effective, environment-friendly, and cost-effective alternatives. These plants, which synthesized green NPs, have demonstrated a remarkable antimicrobial activity against numerous waterborne bacteria such as *E.coli*, *V. cholerae*, *S. typhi*, due to their rich diversity of phytochemicals which are involved in the synthesis process of NPs. Unlike conventional physical and chemical techniques, green synthesis using underutilized plants reduces toxicity risks and aligns with principles of environmental sustainability. Despite the advances in characterization and biosynthesis techniques have strengthened the foundation of this field, several obstacles still exist. These include limited mechanistic understanding of NPs' toxicity and potential environmental bioaccumulation. Additionally, the inconsistent use of biosynthetic protocols and the lack of regulatory frameworks further hinder the translation of these NPs into real-world applications. Future research should concentrate on exploring more underexplored plant species, comprehending the toxicity mechanism of NPs, and creating standardized safety assessments. By bridging current gaps in knowledge and addressing biosafety concerns, plant-based nanotechnology holds immense potential in revolutionizing water treatment and ensuring safer water sources for global usage.

Author's Contribution SS—Writing original draft, data collection, conceptualization; RB—conceptualization, final approval.

Declaration of Competing Interests The authors have no conflicts of interest to declare that are relevant to the content of this chapter.

Ethical Approval Not applicable.

References

1. Cabral JP (2010) Water microbiology. Bacterial pathogens and water. Int J Environ Res Public Health 7(10):3657–3703. https://doi.org/10.3390/ijerph7103657
2. Forstinus NO, Ikechukwu NE, Emenike MP et al (2016) Water and waterborne diseases: a review. Int J Trop Dis Health 12(4):1–14. https://doi.org/10.9734/IJTDH/2016/21895

3. Altammar KA (2023) A review on nanoparticles: characteristics, synthesis, applications, and challenges. Front Microbiol 14:1155622. https://doi.org/10.3389/fmicb.2023.1155622
4. Selmani A, Kovačević D, Bohinc K (2022) Nanoparticles: from synthesis to applications and beyond. Adv Colloid Interf Sci 303:102640. https://doi.org/10.1016/j.cis.2022.102640
5. Sharma S, Dadhwal R, Banerjee R (2025) Nanoparticle assisted phytoremediation: an eco-friendly approach for removal of heavy metals from the environment. J Environ Sci 159:705–720. https://doi.org/10.1016/j.jes.2025.05.034
6. Miu BA, Dinischiotu A (2022) New green approaches in nanoparticles synthesis: an overview. Molecules 2(19):6472. https://doi.org/10.3390/molecules27196472
7. Jin D, Kong X, Cui B et al (2018) Bacterial communities and potential waterborne pathogens within the typical urban surface waters. Sci Rep 8(1):1–9. https://doi.org/10.1038/s41598-018-31706-w
8. Oves M, Arshad M, Khan MS et al (2015) Anti-microbial activity of cobalt doped zinc oxide nanoparticles: targeting water borne bacteria. J Saudi Chem Soc 19(5):581–588. https://doi.org/10.1016/j.jscs.2015.05.003
9. Falinski M, Turley R, Kidd J et al (2020) Doing nano-enabled water treatment right: sustainability considerations from design and research through development and implementation. Environ Sci Nano 7(11):3255–3278. https://doi.org/10.1039/D0EN00584C
10. Olawade DB, Wada OZ, Fapohunda O et al (2024) Nanoparticles for microbial control in water: mechanisms, applications, and ecological implications. Front Nanotechnol 6:1427843. https://doi.org/10.3389/fnano.2024.1427843
11. Primo JDO, Horsth DF, Correa JDS et al (2022) Synthesis and characterization of Ag/ZnO nanoparticles for bacteria disinfection in water. Nanomaterials 12(10):1764. https://doi.org/10.3390/nano12101764
12. Aboyewa JA, Sibuyi NR, Meyer M et al (2021) Green synthesis of metallic nanoparticles using some selected medicinal plants from southern africa and their biological applications. Plants 10(9):1929. https://doi.org/10.3390/plants10091929
13. Kumari S, Sarkar L (2021) A review on nanoparticles: structure, classification, synthesis & applications. J Sci Res 65(8):42–46. https://doi.org/10.37398/JSR.2021.650809
14. Khan S, Hossain MK (2022) Classification and properties of nanoparticles. In: Rangappa SM, Parameswaranpillai J, Gowda TGY, Siengchin S, Seydibeyoglu MO (eds) Nanoparticle-based polymer composites, 1st edn. Elsevier, Amsterdam, pp 15–54
15. Mukherjee P, Ahmad A, Mandal D et al (2001) Fungus-mediated synthesis of silver nanoparticles and their immobilization in the mycelial matrix: a novel biological approach to nanoparticle synthesis. Nano Lett 1(10):515–519. https://doi.org/10.1021/nl0155274
16. Gour A, Jain NK (2019) Advances in green synthesis of nanoparticles. Artif Cells Nanomed Biotechnol 47(1):844–851. https://doi.org/10.1080/21691401.2019.1577878
17. Jadoun S, Arif R, Jangid NK et al (2021) Green synthesis of nanoparticles using plant extracts: a review. Environ Chem Lett 19(1):355–374. https://doi.org/10.1007/s10311-020-01074-x
18. Gangadoo S, Taylor-Robinson A, Chapman J (2015) Nanoparticle and biomaterial characterisation techniques. Mater Technol 30(sup5):B44–B56. https://doi.org/10.1179/1753555714Y.0000000201
19. Padulosi S, Hoeschle-Zeledon I (2004) Underutilized plant species: what are they? Leisa-Leusden 20:5–6
20. Nirmala C, Bajwa HK, Oinam S (2024) Bamboo mediated green synthesis of silver nanoparticles-A new approach towards utilization of an underutilized plant. Adv Bamboo Sci 6:100061. https://doi.org/10.1016/j.bamboo.2024.100061
21. Chokkareddy R, Redhi GG (2018) Green synthesis of metal nanoparticles and its reaction mechanisms. In: Kanchi S, Ahmed S (eds) Green metal nanoparticles: synthesis, characterization and their applications. Wiley, USA, pp 113–139
22. Malik P, Shankar R, Malik V et al (2014) Green chemistry based benign routes for nanoparticle synthesis. J Nanopart 2014(1):302429. https://doi.org/10.1155/2014/302429

23. Mutukwa D, Taziwa R, Khotseng LE (2022) A review of the green synthesis of ZnO nanoparticles utilising Southern African indigenous medicinal plants. Nano 12(19):3456
24. Okigbo R, Anyaegbu C (2021) Underutilized plants of Africa. J Biol Nat 13(2):34–49. https://doi.org/10.3390/nano12193456
25. Haleem A, Javaid M, Singh RP et al (2023) Applications of nanotechnology in medical field: a brief review. Glob Health J 7(2):70–77. https://doi.org/10.1016/j.glohj.2023.02.008
26. Chakraborty A, Haque SM, Dey D et al (2022) Phytogenic silver nanoparticles from tissue-cultured *Kaempferia angustifolia*—an underutilized medicinal herb: a comparative antibacterial study on urinary pathogens. J Genet Eng Biotechnol 20(1):131. https://doi.org/10.1186/s43141-022-00414-4
27. Banerjee S, Islam S, Chattopadhyay A et al (2022) Synthesis of silver nanoparticles using underutilized fruit *Baccaurea ramiflora* (Latka) juice and its biological and cytotoxic efficacy against MCF-7 and MDA-MB 231 cancer cell lines. S Afr J Bot 145:228–235. https://doi.org/10.1016/j.sajb.2021.09.016
28. Sahu PP, Behera L, Nayak S et al (2020) Health benefits of Jamun (*Syzygium cumini*) an underutilised fruit: a ray in nanotechnology field. J Pharmacogn Phytochem 9(5):74–80
29. Labulo AH, Adesuji ET, Dedeke OA et al (2016) A dual-purpose silver nanoparticles biosynthesized using aqueous leaf extract of *Detarium microcarpum*: an under-utilized species. Talanta 160:735–744. https://doi.org/10.1016/j.talanta.2016.07.066
30. Odeniyi MA, Adegbolagun TI, Odeniyi OA et al (2025) Oral dissolving films of native and modified starches from underutilized crops as delivery vehicles for green synthesized *Bridelia ferruginea*-zinc oxide nanoparticles. Next Res 2(3):100502. https://doi.org/10.1016/j.nexres.2025.100502
31. Dakshanamurthy D, Govindarajan VU, Muthuraman MS (2023) Facile preparation of *Crescentia cujete*–mediated copper oxide nanoparticles for dye degradation properties. Biomass Convers Biorefinery 15:1–12. https://doi.org/10.1007/s13399-023-04940-0
32. Baghani M, Es-Haghi A (2020) Characterization of silver nanoparticles biosynthesized using *Amaranthus cruentus*. Bioinspired Biomimetic Nanobiomater 9(3):129–136. https://doi.org/10.1680/jbibn.18.00051
33. Fabiyi O, Bello T, Lateef A et al (2024) Evaluation of *Borreria verticillata* (L.) synthesized copper, iron and zinc nanoparticles against *Meloidogyne Incognita* on cowpea and the effect on cowpea seed quality. Journal of Crop Health 76(6):1507–1518. https://doi.org/10.1007/s10343-024-01036-1
34. Al Rashdi BM, Germoush MO, Sani SS et al (2023) Biosynthesis of *Salvia hispanica* based silver nanoparticles and evaluation of their antibacterial activity *in-vitro* and rat model. Pak Vet J 43(2):283–289. https://doi.org/10.29261/pakvetj/2023.035
35. Singh K, Yadav S (2025) Biosynthesis of a range of ZnO nanoparticles utilising *Salvia hispanica* L. seed extract and evaluation of their bioactivity. Sci Rep 15(1):043. https://doi.org/10.1038/s41598-025-87355-3
36. Devanesan S, David HA, Ranjitsingh AJ et al (2024) Efficient biogenesis of calcium oxide nanoparticles using the extract of *Eleusine coracana* seeds and their application against multidrug-resistant ocular bacterial pathogens. Environ Res 251:118632. https://doi.org/10.1016/j.envres.2024.118632
37. Almudhafar SM, Al-Hamdani MA (2022) Antibacterial and anticancer effects of silver nanoparticles synthesised using *Eragrostis tef* and *Vitellaria paradoxa* seeds extract. Basrah J Agric Sci 35(2):132–159. https://doi.org/10.37077/25200860.2022.35.2.10
38. Adeyemi JO, Onwudiwe DC, Oyedeji AO (2022) Biogenic synthesis of CuO, ZnO, and CuO–ZnO nanoparticles using leaf extracts of *Dovyalis caffra* and their biological properties. Molecules 27(10):3206. https://doi.org/10.3390/molecules27103206
39. Al-Rajhi AM, Salem SS, Alharbi AA et al (2022) Ecofriendly synthesis of silver nanoparticles using Kei-apple (*Dovyalis caffra*) fruit and their efficacy against cancer cells and clinical pathogenic microorganisms. Arab J Chem 15(7):103927. https://doi.org/10.1016/j.arabjc.2022.103927

40. Majoumouo MS, Sibuyi NRS, Tincho MB et al (2019) Enhanced anti-bacterial activity of biogenic silver nanoparticles synthesized from *Terminalia mantaly* extracts. Int J Nanomedicine 14:9031–9046. https://doi.org/10.2147/IJN.S223447
41. Lavanya B, Aparna Y, Ramana M (2023) Synthesis, characterization of zinc oxide nanoparticles using *Terminalia mantaly* leaf extract and their antibacterial activity. Mater Today Proc 92:1011–1016. https://doi.org/10.1016/j.matpr.2023.05.001
42. Suresh D, Nethravathi P, Rajanaika H et al (2015) Green synthesis of multifunctional zinc oxide (ZnO) nanoparticles using *Cassia fistula* plant extract and their photodegradative, antioxidant and antibacterial activities. Mater Sci Semicond Process 31:446–454. https://doi.org/10.1016/j.mssp.2014.12.023
43. Adebayo IA, Arsad H, Gagman HA et al (2020) Inhibitory effect of eco-friendly naturally synthesized silver nanoparticles from the leaf extract of medicinal *Detarium microcarpum* plant on pancreatic and cervical cancer cells. Asian Pac J Cancer Prev: APJCP 21(5):1247. https://doi.org/10.31557/APJCP.2020.21.5.1247
44. Malanovic N, Lohner K (2016) Gram-positive bacterial cell envelopes: the impact on the activity of antimicrobial peptides. Biochim Biophys Acta (BBA)-Biomembr 1858(5):936–946. https://doi.org/10.1016/j.bbamem.2015.11.004
45. Huang KC, Mukhopadhyay R, Wen B et al (2008) Cell shape and cell-wall organization in gram-negative bacteria. Proc Natl Acad Sci 105(49):19282–19287. https://doi.org/10.1073/pnas.080530910
46. Franco D, Calabrese G, Guglielmino SPP et al (2022) Metal-based nanoparticles: antibacterial mechanisms and biomedical application. Microorganisms 10(9):1778. https://doi.org/10.3390/microorganisms10091778
47. Deshmukh RA, Joshi K, Bhand S et al (2016) Recent developments in detection and enumeration of waterborne bacteria: a retrospective minireview. MicrobiologyOpen 5(6):901–922. https://doi.org/10.1002/mbo3.383
48. Theron J, Cloete T (2002) Emerging waterborne infections: contributing factors, agents, and detection tools. Crit Rev Microbiol 28(1):1–26. https://doi.org/10.1080/1040-840291046669
49. Hunter PR (2003) Drinking water and diarrhoeal disease due to *Escherichia coli*. J Water Health 1(2):65–72. https://doi.org/10.2166/wh.2003.0008
50. Kumar D, Kumari S (2013) *Klebsiella*: In drinking water. Int J Pharm Sci Invent 2(12):38–42
51. Mena KD, Gerba CP (2009) Risk assessment of *Pseudomonas aeruginosa* in water. Rev Environ Contam Toxicol 201:71–115. https://doi.org/10.1007/978-1-4419-0032-6_3
52. Ghalehnoo ZR (2018) Diseases caused by *Staphylococcus aureus*. Int J Med Health Res 4(11):65–67
53. Rodrigues GR, Lopez-Abarrategui C, de la Serna Gómez I et al (2019) Antimicrobial magnetic nanoparticles based-therapies for controlling infectious diseases. Int J Pharm 555:356–367. https://doi.org/10.1016/j.ijpharm.2018.11.043
54. Viganor L, Howe O, McCarron P et al (2017) The antibacterial activity of metal complexes containing 1, 10-phenanthroline: potential as alternative therapeutics in the era of antibiotic resistance. Curr Top Med Chem 17(11):1280–1302
55. Mukha IP, Eremenko A, Smirnova N et al (2013) Antimicrobial activity of stable silver nanoparticles of a certain size. Appl Biochem Microbiol 49(2):199–206. https://doi.org/10.1134/S0003683813020117
56. Gu X, Xu Z, Gu L et al (2021) Preparation and antibacterial properties of gold nanoparticles: a review. Environ Chem Lett 19(1):167–187. https://doi.org/10.1007/s10311-020-01071-0
57. Saliani M, Jalal R, Goharshadi EK (2015) Effects of pH and temperature on antibacterial activity of zinc oxide nanofluid against *Escherichia coli* O157: H7 and *Staphylococcus aureus*. Jundishapur J Microbiol 8(2):e17115. https://doi.org/10.5812/jjm.17115
58. Ramalingam B, Parandhaman T, Das SK (2016) Antibacterial effects of biosynthesized silver nanoparticles on surface ultrastructure and nanomechanical properties of Gram-negative bacteria viz. *Escherichia coli* and *Pseudomonas aeruginosa*. ACS Appl Mater Interfaces 8(7):4963–4976. https://doi.org/10.1021/acsami.6b00161

59. Allawadhi P, Singh V, Khurana A et al (2021) Silver nanoparticle based multifunctional approach for combating COVID-19. Sens Int 2:100101. https://doi.org/10.1016/j.sintl.2021.100101
60. Krishnaraj C, Jagan E, Rajasekar S et al (2010) Synthesis of silver nanoparticles using *Acalypha indica* leaf extracts and its antibacterial activity against water borne pathogens. Colloids Surf B Biointerfaces 76(1):50–56. https://doi.org/10.1016/j.colsurfb.2009.10.008
61. Ali HR, Emam AN, Koraney NF et al (2020) Combating the prevalence of water-borne bacterial pathogens using anisotropic structures of silver nanoparticles. J Nanopart Res 22(2):47. https://doi.org/10.1007/s11051-020-4760-6
62. Bernardo WLDC, Boriollo MFG, Tonon CC et al (2022) Biosynthesis of silver nanoparticles from *Syzygium cumini* leaves and their potential effects on odontogenic pathogens and biofilms. Front Microbiol 13:995521. https://doi.org/10.3389/fmicb.2022.995521
63. Kumar R, Umar A, Kumar G et al (2017) Antimicrobial properties of ZnO nanomaterials: a review. Ceram Int 43(5):3940–3961. https://doi.org/10.1016/j.ceramint.2016.12.062
64. Jiang S, Lin K, Cai M (2020) ZnO nanomaterials: current advancements in antibacterial mechanisms and applications. Front Chem 8:580. https://doi.org/10.3389/fchem.2020.00580
65. Ma X, Zhou S, Xu X et al (2022) Copper-containing nanoparticles: mechanism of antimicrobial effect and application in dentistry-a narrative review. Front Surg 9:905892. https://doi.org/10.3389/fsurg.2022.905892
66. Khatri FA, Solangi IB, Memon GZ et al (2024) Antibacterial application of copper nanoparticles biosynthesized by water caltrop pod. J Appl Res Plant Sci 5(02):288–296. https://doi.org/10.38211/joarps.2024.05.282
67. Mobed A, Hasanzadeh M, Seidi F (2021) Anti-bacterial activity of gold nanocomposites as a new nanomaterial weapon to combat photogenic agents: recent advances and challenges. RSC Adv 11(55):4688–34698. https://doi.org/10.1039/D1RA06030A
68. Bystrzejewska-Piotrowska G, Golimowski J, Urban PL (2009) Nanoparticles: their potential toxicity, waste and environmental management. Waste Manag 29(9):2587–2595. https://doi.org/10.1016/j.wasman.2009.04.001
69. Gupta R, Xie H (2018) Nanoparticles in daily life: applications, toxicity and regulations. J Environ Pathol Toxicol Oncol 37(3):209–230. https://doi.org/10.1615/JEnvironPatholToxicolOncol.2018026009
70. Ettlinger R, Lächelt U, Gref R et al (2022) Toxicity of metal–organic framework nanoparticles: from essential analyses to potential applications. Chem Soc Rev 51(2):464–484. https://doi.org/10.1039/D1CS00918D
71. Anand U, Carpena M, Kowalska-Góralska M et al (2022) Safer plant-based nanoparticles for combating antibiotic resistance in bacteria: a comprehensive review on its potential applications, recent advances, and future perspective. Sci Total Environ 82:153472. https://doi.org/10.1016/j.scitotenv.2022.153472

Chapter 6
Underutilized Plant-Derived Nanoparticles Against Gram-Negative and Gram-Positive Bacteria

Indu Bhardwaj, Shriyanshu Thakur, Nidhi Bhardwaj, Kaustubh Chandrakant Khaire, and Habtamu Mekonnen

Abstract Antimicrobial resistance (AMR) is a growing worldwide health crisis that undermines the efficacy of traditional antibiotics against Gram-negative and Gram-positive bacteria, including multidrug-resistant bacteria like methicillin-resistant *Staphylococcus aureus* (MRSA). Nanotechnology presents potential solutions to AMR in the form of plant-derived nanoparticles (PDNPs), which are biocompatible, environmentally friendly to synthesize, and highly effective antibacterial agents. Underexploited plants like *Bambusa* spp., *Amaranthus* spp., and *Pulicaria* spp. offer a rich source of phytochemicals, such as flavonoids, alkaloids, terpenoids, and phenolics—that act as natural reducing and stabilizing agents in green nanoparticle synthesis, as well as increasing nanoparticle stability and antibacterial activity. PDNPs have their antibacterial action by several mechanisms such as bacterial membrane disruption, production of reactive oxygen species (ROS), interference with biofilm formation, and interference with DNA and protein synthesis, thus decreasing the risk of resistance. Green synthesis with underutilized plants is economical, sustainable, and eco-friendly as compared to traditional chemical and physical routes. Though promising, there are issues like inconsistency in phytochemical content, lack of extensive toxicological assessment, scalability, and regulatory hurdles that must be overcome for clinical applicability. This chapter

I. Bhardwaj (✉) · S. Thakur (✉)
Division of Microbiology, Career Point University, Hamirpur, Himachal Pradesh, India

N. Bhardwaj
Center of Advanced Innovation Technologies, VŠB-Technical University of Ostrava, Ostrava, Poruba, Czech Republic

Department of Chemistry, Faculty of Science, University of Hradec Kralove, Hradec Kralove, Czech Republic

K. C. Khaire
Department of Chemical Engineering, Faculty of Science and Engineering, The University of Manchester, Manchester, UK

H. Mekonnen
Department of Biology, Bahir Dar University, Bahir Dar, Ethiopia

© The Author(s), under exclusive license to Springer Nature Switzerland AG 2026
E. Nepovimova et al. (eds.), *Underutilized Plants Based-Nanoparticles: Herbal Remedy to Treat Human Diseases*, https://doi.org/10.1007/978-3-032-20384-7_6

contains an extensive review of the phytochemistry of underutilized crops, PDNP characterization, antibacterial mechanisms of PDNPs, opportunities, and challenges in the development of these nanoparticles as next-generation antibacterial agents.

Graphical Abstract

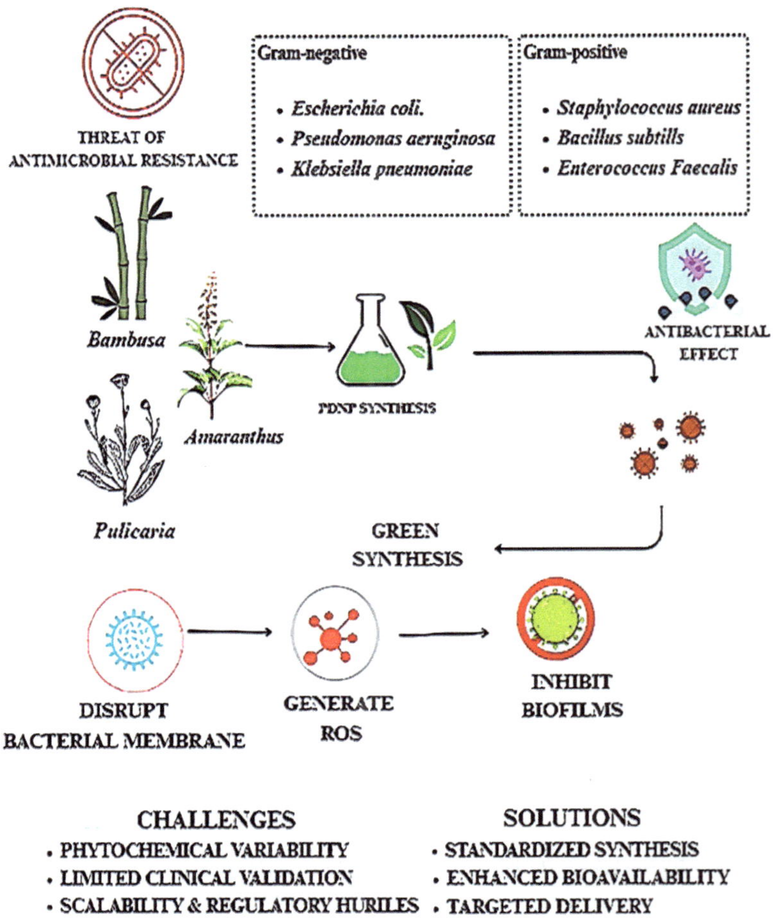

Keywords Antibacterial activity · Bacteria · Green synthesis · PDNPs · Underutilized plants

6.1 Introduction

6.1.1 Antimicrobial Resistance: A Global Health Crisis

Antimicrobial resistance (AMR) has been identified as a significant global health concern with resistant pathogens propagating extended infections, escalating healthcare expenses, and higher death rates [1]. The WHO estimates that AMR could account for up to ten million deaths each year by 2050 if new interventions are not established [2]. Gram-negative bacteria such as *Escherichia coli*, *Klebsiella pneumoniae*, and *Pseudomonas aeruginosa*, through their impermeable membranes and efflux pumps, are intrinsically resistant, whereas Gram-positive pathogens such as *Staphylococcus aureus* and *Enterococcus faecalis* [3] have developed resistance against last-resort antibiotics such as methicillin and vancomycin [4]. This grim crisis also suggests developing new antibacterial solutions capable of countering multidrug-resistant pathogens.

6.1.2 Nanotechnology and Plant-Derived Nanoparticles

Nanotechnology offers a new approach in fighting AMR, while the biocompatibility and eco-friendly synthesis of plant-derived nanoparticles (PDNPs) make this subclass unique. Metal or metal oxide nanoparticles such as silver, gold, zinc oxide, and copper oxide exhibit good antibacterial activity due to their nanometric size [5], large surface-to-volume ratio, and ability to disrupt microbial membranes, generate ROS, interfere with intracellular molecules, and inhibit biofilm formation. As opposed to the traditional expensive chemical syntheses, which are hazardous to the environment, green synthesis using plant extracts relies on phytochemicals as reducing and stabilizing agents for minimal impact on the environment and enhanced biocompatibility [5, 6]. This is even more relevant for PDNPs based on plant extracts that have been neglected as manufacturers have not yet realized the benefits of exploiting abundant and diverse bioactive metabolites, which can impact nanoparticle morphology and antimicrobial activity.

6.1.3 Underutilized Plants and Their Antimicrobial Potential

Neglected crops are often not considered in mainstream agricultural or biomedical research. They are highly diverse plants with a huge number of phytochemicals that have strong antibacterial activity against Gram-negative and Gram-positive pathogens [7]. Most of these crops have been traditionally used for an extensive time in Indigenous healing systems across Africa, Asia, or the Middle East, focusing on their broad-spectrum medicinal properties. Their antibacterial activity is believed to be attributed to high content of secondary metabolites, including alkaloids, flavonoids, terpenoids, and phenolic compounds, many are documented to show varying antibacterial activity [8]. For example, phenolic components and lignin derivatives from *Bambusa vulgaris* inhibit *S. aureus* and *E. coli*. *Amaranthus* species carry a unique profile of flavonoids and saponins with bioactivity against infectious agents such as *P. aeruginosa* and *Bacillus subtilis* [9, 10]. *Pulicaria dysenterica* is another neglected crop whose terpenoids and sesquiterpenes are already known to be active against *E. coli* and *S. aureus* [11]. Finally, these neglected crops investigated herein provide more ecological and cultural advantages by growing on unproductive land with very minimal input, hence opening an avenue toward sustainable social and environmental sources of new antibacterial agents [12].

However, their application in direct form is challenging because of various difficulties, which include the geographic and climatic variability of the phytochemical content, further leading to variability in efficacy, besides issues related to standardization [13]. There are limited in vivo and clinical trials across these applications that may deter their integration into conventional medicine. Besides, ecological depletions due to over-harvesting and a non-standardize extraction process restrict scalability [10, 14]. Considering such constraints, such challenges are being overcome by more sophisticated methods of nanotechnology to enhance stability, reproducibility, and efficacy towards successfully producing PDNPs as antibacterial agents.

6.1.4 The Unleashed Potential of Underutilized Plants in Nanotechnology

PDNPs synthesized from previously underutilized plants may be a way to treat AMR by bypassing the limitations associated with direct plant antimicrobials. Many phytochemicals, such as flavonoids, terpenoids, and alkynes, can serve as reducing and capping agents in the formation of metal nanoparticles and will dictate their size, shape, and antibacterial activity [15]. Silver nanoparticles made using these plants have a significant effect against *E. coli*, *S. aureus*, and *P. aeruginosa*, including methicillin-resistant *S. aureus* (MRSA) by lysing bacterial cell walls and inducing ROS [16, 17]. Other previously underutilized species, such as *Embelica officinalis* [18] and *Syzygium cumini* [19], also produce PDNPs that increase antibacterial activity against resistant organisms. Nanotechnology circumvents the major drawbacks of phytochemical variability and scalability by developing improved reproducibility and targeted delivery, befitting the sustainable development goals. However, it also bears certain limitations, such as the absence of standard protocols for synthesizing those nanoparticles and very few in vivo studies; therefore, it needs efforts from multidisciplinary research to bring these nanomaterials at clinical aspects. By leveraging the phytochemical repertoire of such plants [14], PDNPs offer unique, biologically inspired solutions to the AMR emergency.

In this chapter, we analyze the AMR crisis, nanotechnology potential with emphasis on PDNPs, and antibacterial potential of underexploited plants, putting into focus *B. vulgaris*, *Amaranthus* spp., and *Pulicaria* spp. We set a stage for subsequent chapters that explain their mechanisms, incorporation into nanotechnology, and how to combat multidrug-resistant bacterial diseases by focusing on their diversity in phytochemicals, inadequacies, and how the use of nanotechnology solves them.

6.2 Phytochemical-Driven Antibacterial Modes of Underutilized Plants

The underutilized plants like *B. vulgaris*, *Amaranthus* spp., and *P. dysenterica* are rich in phytochemicals such as flavonoids, alkaloids, terpenoids, and phenolics that possess potent antibacterial action against Gram-negative and Gram-positive bacteria. The compounds inhibit through disruption of membranes, inhibition of biofilms, the production of reactive oxygen species, and inhibition of crucial cell processes. Their activity via multi-target mechanisms increases the therapeutic potential and supports green drug discovery. The existence of such phytochemical richness indicates their potential for nanotechnology integration in fighting antimicrobial resistance with success [20].

6.2.1 Phytochemicals with Antibacterial Activity

The antibacterial action of underexploited plants is derived from secondary metabolites acting on Gram-negative and Gram-positive bacteria through more than one mechanism of action. The major classes include:

(a) *Alkaloids:* Nitrogenous alkaloids, such as berberine, present in *Berberis* spp. (Indian barberry), disrupt microbial cell membranes and inhibit DNA replication and protein synthesis, thus showing effectiveness against pathogenic organisms like *S. aureus*, *E. coli*, and *K. pneumoniae* [21].

(b) *Flavonoids:* This compound found in *Moringa oleifera* (drumstick tree) and *Amaranthus* spp., quercetin, inhibits bacterial enzyme activity, alters cell wall stability, and inhibits microbial adhesion, acting against *E. coli* and *P. aeruginosa* [22].

(c) *Terpenoids:* Found in *Ocimum gratissimum* essential oils (African basil) [23], terpenoids such as eugenol and thymol weaken microbial membranes and interfere with quorum sensing [24, 25], an essential process in bacterial virulence and biofilm development.

(d) *Polyphenolic Compounds:* Polyphenols, including gallic acid from *Punica granatum* (pomegranate) and lignin derivatives from *Bambusa* spp., interfere with microbial enzyme function, inhibit adhesion, and trigger oxidative stress, interfering with pathogens such as *K. pneumoniae* [26].

(e) *Saponins:* They cause lysis in bacteria by damaging the integrity of the membrane, and are active against *B. cereus*, *S. aureus*, and *P. aeruginosa* [27].

6.2.2 Example of Underutilized Plants with Potential Antibacterial Activity

Table 6.1 summarizes eight underutilized plant species, focusing on *B. vulgaris*, *Amaranthus* spp., and *Pulicaria* spp., and their bioactive compounds, antibacterial activity against Gram-negative and Gram-positive bacteria, and traditional use.

6.2.3 Mechanisms and Role in the Fight Against Antimicrobial Resistance

The antimicrobial properties of these underutilized plant species arise from various pathways enhancing their effectiveness against multidrug-resistant bacteria [40]. Phytochemicals such as phenols, saponins, and terpenoids act by compromising bacterial cell membrane stability which leads to lysis, for instance, lignin derivatives in *B. vulgaris* exhibit strong activity against both *S. aureus* and *E. coli*, while *Pulicaria* species also contain sesquiterpenes with similar antibacterial activity [40]. Alkaloids and flavonoids interfere with bacterial protein production by disrupting metabolism, enzymatic activity, and efflux system mechanisms and decreasing the viability of the drug-resistant pathogens such as *S. aureus* and *E. coli* [41], whereas certain plant-derived compounds like berberine inhibited nucleic acid synthesis by directly inhibiting DNA gyrase, a critical enzyme for DNA replication in *E. coli* and *S. aureus* [42]. Terpenoids and flavonoids from some species disrupted quorum-sensing signaling and biofilm formation of *P. aeruginosa* [43].

Moreover, phenolic and flavonoid compounds create oxidative stress by enhancing reactive oxygen species formation that disrupts bacterial DNA, proteins, and lipids [44]. Some of these phytoconstituents also exhibit synergistic relationships with non-phytochemical antibiotics. For instance, *S. cumini* extracts have been shown to enhance the activity of ampicillin against *E. coli* and *S. aureus* and reduce the effects of multidrug-resistant strains [36]. Taken all together, the provocation of these different mechanisms and activities makes these seldom useful plants viable candidates to address the problem of antimicrobial resistance as they have broad-spectrum activity against drug-resistant bacteria such as MRSA and multidrug-resistant *E. coli*. In addition, the extent of multitargeted activities reduces the opportunity for development of resistance, and their traditional medicinal usage, ethnobotany, and resilience in suboptimal environments would meet sustainable development goals and thus position them as environmentally sustainable therapeutic resources [45, 46].

Table 6.1 Underutilized plants with antibacterial potential

Plant name	Scientific name	Active compounds	Antibacterial effects	Traditional uses	References
Bamboo	*B. vulgaris*	Phenolics, flavonoids, lignin	Inhibits *E. coli*, *S. aureus*	Wound healing and injuries in the skin	[28]
Amaranth	*Amaranthus* spp.	Flavonoids, saponins, phenolic acids	Against *B. subtilis*, *P. aeruginosa*	Wound treatment, infection prevention in African and Latin American cultures	[29]
Fleabane	*Pulicaria* spp.	Terpenoids, flavonoids, sesquiterpenes	Inhibits *S. aureus*, *E. coli*	Skin and respiratory infections in Middle Eastern traditional medicine	[30]
Drumstick tree	*M. oleifera*	Flavonoids, glucosinolates, phenolic acids	Inhibits *S. aureus*, *E. coli*	Wound healing, controlling infection in African and South Asian medicine	[31, 32]
Neem	*Azadirachta indica*	Azadirachtin, nimbidin, limonoids	Inhibits *S. aureus* and *P. aeruginosa*	Oral hygiene and skin-related infections in Ayurveda	[33, 34]
Guduchi	*Tinospora cordifolia* (Underutilized plant from Uttar Pradesh, India)	Berberine, palmatine, glycosides	Inhibits *S. typhi*, *B. subtilis*	Fevers, infections in Ayurveda	[35]
Jamun	*S. cumini*	Alkaloids, flavonoids, tannins, saponins, steroids, terpenoids, phenols, glycosides, resins	Effective against MRSA and *E. coli*; synergistic with ampicillin	Strengthen teeth and gums; treat diabetes, constipation, fever, gastropathy; inhibit blood discharge in feces	[36]
African Basil	*O. gratissimum*	Eugenol, thymol, linalool	Inhibits *E. faecalis*, *S. enterica*	Respiratory infections, wound care in African ethnomedicine	[37–39]

6.2.4 Translational Opportunities of Underutilized Plants

Underutilized plants offer a unique opportunity to develop next-generation antibiotics based on nanotechnology. In addition to providing the phytochemicals for green synthesis of nanoparticles, they are also enhancing the stability and therapeutic activity of nanoparticles [47]. The ability to deliver this type of material, with minimal systemic toxicity, is now becoming apparent with the incorporation of PDNPs into wound dressings, coatings, and oral products [48]. In addition, cultivating robust species such as *Amaranthus* and *Pulicaria* spp. in marginal areas contributes to sustainable agriculture and rural development [12]. The potential in veterinary medicine, food preservation, and environmental disinfection also extends the public health impact of PDNPs on AMR [49]. This potential suggests a translational capability of underutilized plants towards scalable, sustainable, and diverse antimicrobials for the global threat of AMR.

6.3 Underutilized Plant Nanotechnology-Aided Antibacterial Alternatives

The growing burden of AMR has made conventional antibiotic use less effective; therefore, the need for new therapeutic approaches has become pressing [50]. Nanotechnology is a very versatile platform for designing nanoscale products with specific physicochemical properties and sophisticated modes of antibacterial action. This approach ensures increased antibacterial activity with underexploited plants, with the additional possibility of green and sustainable production [14, 51]. PDNPs, therefore, represent a novel approach to tackling one of the biggest challenges facing modern medicine.

6.3.1 Nanotechnology and Its Role in Combating AMR

Nanoparticles capable of interacting with microbial cells at a molecular level have emerged through the utilization of matter between 1 and 100 nm in size, as allowed by nanotechnology. PDNPs show broad-spectrum action against Gram-negative bacteria such as *E. coli*, *P. aeruginosa*, and *K. pneumoniae*, and Gram-positive bacteria including *S. aureus*, *B. subtilis*, and *E. faecalis*, with remarkable activity against MRSA [52]. Unlike the conventional antibiotics, which tend to strike one pathway, nanoparticles act multimodally: membrane disruption, induction of ROS, biofilm inhibition, and disturbance of DNA and protein synthesis, thus minimizing the chances of resistance [53].

6.3.2 *Types and Classification of Nanoparticles*

The synthesis of NPs for antibacterial applications can be done through a variety of synthesis pathways [54]:

6.3.2.1 Modes of Synthesis Strategies

There are two main approaches used for the production of NPs:

A. *Top-down Method*
The top-down approach, such as milling and laser ablation, yields very pure metallic nanoparticles but is costly and less scalable [55, 56].
B. *Bottom-down Approach*
Unlike that, bottom-up procedures, especially green synthesis, use plant extracts to decrease and stabilize metal ions such as silver (Ag^{+}) and gold (Au^{3+}) to form stable, environmentally friendly nanoparticles with high antibacterial activities [57, 58].

6.3.2.2 Categorization of Nanoparticles

Antibacterial nanoparticles are divided into three wide categories: metal, metal oxides, and organic [59].

A. *Metallic Nanoparticles (MtNPs)*
MtNPs, particularly silver, gold, and copper, are broad-spectrum active. For instance, silver nanoparticles penetrate the bacterial cell wall, release ions, and interfere with DNA [17]. Flavonoid-stabilized gold nanoparticles improve the antibiotic action against *K. pneumoniae* [60, 61], whereas copper nanoparticles mostly cause oxidative stress [62].
B. *Metal Oxide Nanoparticles (MONPs)*
MONPs such as zinc oxide (ZnO) and copper oxide (CuO), interfere with enzymatic function and produce ROS, with ZnO having good biofilm inhibition against *P. aeruginosa* [63, 64].
C. *Organic Nanoparticles (ONPS)*
ONPs, such as chitosan-based systems, are efficient at quorum-sensing disruption in pathogens like *K. pneumoniae* and *E. faecalis*. Furthermore, lipid-based carriers enhance the biofilm penetration, which in turn enhances their antibacterial activity [65, 66].

6.3.3 Integration of Underutilized Plants with Nanotechnology

Underexploited species yield phytochemicals that aid in nanoparticle synthesis while enhancing antibacterial activity [67]. *B. vulgaris* phenolic and lignin derivatives reduce silver ions into nanoparticles that inhibit *E. coli* and *S. aureus*. Flavonoids of *Amaranthus* spp. stabilize ZnO nanoparticles, which increase oxidative stress activity towards *P. aeruginosa*. Terpenoids and sesquiterpenes from *Pulicaria* spp. control copper nanoparticles against MRSA [68]. Nanotechnology also normalizes variation in phytochemical composition, enhancing reproducibility and scalability [69]. The green synthesis methods are done in such a way that it ensures clinical applicability and environmental safety. Furthermore, in vivo studies have started validating efficacy against drug-resistant strains [70].

6.3.4 Green Synthesis Using Underutilized Plants

The biological/or green synthesis from plant extracts does not contain toxic reagents, and the phytochemicals act as natural capping and reducing agents [71]. Underutilized crops are still unexploited but contain metabolites that yield stable, efficient nanoparticles with excellent antibacterial activity. The process fits into sustainability objectives by giving scalable, low-cost, and environmentally friendly approaches toward AMR control [72, 73].

6.4 Underutilized Plants' Mediated NPs Against Bacteria

The rapid upsurge of drug-resistant infections necessitates the need to search for new antimicrobial agents. Nanotechnology, green synthesis of PDNPs in particular, is a sustainable, scale-up, and cost-effective means of achieving it [74]. Underutilized plants like species of *Bambusa*, *Amaranthus*, and *Pulicaria* contain many phytochemicals with reducing and capping capabilities, hence facilitating nanoparticle synthesis while furthering antibacterial activity [75]. These PDNPs also showed high activity against Gram-positive and Gram-negative pathogens and were good candidates for sustainable development of antimicrobials.

6.4.1 Phytochemistry Basis in Certain Underutilized Plants

During the biosynthesis of NPs, secondary metabolites such as flavonoids, phenolics, terpenoids, alkaloids, saponins, and betalains play an important role. These inhibit aggregation and transfer electrons to reduce metal ions such as Ag^+ and Zn^{2+} thereby offering a stable and bioactive nanostructure [76].

(a) *Bamboo Species (B. arundinacea, B. vulgaris)*
Bamboo extracts are rich in flavonoids, polysaccharides, and phenolic acids, which can efficaciously reduce silver and zinc ions. For instance, silver nanoparticles prepared using *B. vulgaris* leaf extract exhibited antibacterial activity against Gram-positive and Gram-negative bacteria and photocatalytic dye degradation [74]. ZnO nanoparticles prepared from *B. arundinacea* exerted ROS-mediated bactericidal effects, representing again a double contribution of bamboo to nanoparticle preparation and activity [77, 78].

(b) *Amarnathus Species (A. viridis, A. gangeticus, A. dubius)*
Amaranthus species have betalains, flavonoids, tannins, and phenolic acids that are readily available to interact with metal ions. Betalains possess high reducing power, thereby efficiently converting silver and zinc precursors to nanoparticles. AgNPs from *A. viridis* have the ability to inhibit *S. aureus* and *E. coli* [79], whereas *A. gangeticus* extracts produce nanoparticles with high antioxidant and antibacterial activity [80]. Likewise, *A. dubius* synthesizes AgNPs with remarkable activity against *B. cereus* and *Salmonella* spp., pointing to their low-cost prospect as green nanofactories [81].

(c) *Pulicaria* (*P. undulata, P. Jaubertii* and *P. vulgaris*)
Different species of *Pulicaria* yield flavonoids, phenolics, terpenoids, and sesquiterpene lactones serving as stabilizers and electron donors [82]. *P. vulgaris* aerial part silver nanoparticles exhibit excellent antibacterial and antioxidant activities attributed to the phytochemical capping layers [68]. The extracts of *P. jaubertii* and *P. undulata* also enhance the properties of TiO_2 and AgCl NPs [82–84], hence showing high activity against resistant bacterial strains.

6.4.2 Antimicrobial Activities Against Gram-Negative and Gram-Positive Bacteria

PDNPs made from underutilized crops have shown strong antibacterial effects against both Gram-positive and Gram-negative bacteria. For example, AgNPs, which are silver nanoparticles biosynthesized from bamboo and have been reported to cause significant inhibition in bacterial cell growth as revealed by the decreased number of bacterial cell count and clear inhibition zones formed against *S. aureus* and *B. subtilis* [85]. *Amaranthus*-based nanoparticles also exhibit significant antibacterial activity, with ZnO and AgNPs displaying broad coverage against Gram-positive bacteria. Gram-negative bacteria such as *E. coli* and *P. aeruginosa*—usually more difficult to treat because of their outer membrane—are also sensitive to PDNPs produced from *Bambusa*, *Amaranthus*, and *Pulicaria* species. Postulated mechanisms include increased permeability of the membrane, ROS generation, and bacterial protein and DNA interaction [81] as shown in Fig. 6.1. The antibacterial activity of the nanoparticles isolated from under-exploited plants is shown in Table 6.2.

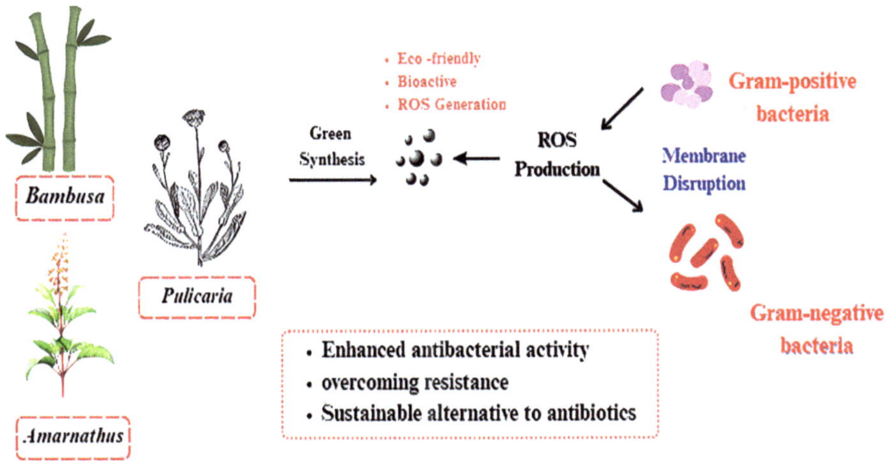

Fig. 6.1 Underutilized PDNPs against Gram-positive and Gram-negative bacteria

Table 6.2 Antibacterial activity of nanoparticles synthesized from underutilized plants

Underutilized plant	Species	Capping agent	Shape	Size	Gram +ve	Gram −ve	References
Bambusa	*B. arundinacea*	Leaf extract	Spherical	20–80 nm	*B. subtilis, S. aureus*	*E. coli, P. aeruginosa*	[78]
	B. vulgaris	Leaves	Spherical	15–78 nm	*S. aureus*	*E. coli*	[85]
	B. arundinacea	Leaf extract	Spherical	20–90 nm	*B. subtilis, S. aureus*	*P. aeruginosa*	[77]
Amaranthus	*A. viridis*	Leaf extract	Spherical	5–20 nm	*B. subtilis, S. aureus*	*E. coli, P. putida*	[79]
	A. dubius	Leaf extract	Spherical	14 nm	*B. cereus, S. aureus*	*E. coli, Salmonella spp*	[81]
	A. gangeticus Linn	Leaf extract	Globular	11–15 nm	*B. subtilis*	*Shigella flexneri*	[80]
Pulicaria	*P. undulata*	Aerial part extract	Spherical	70–80 nm	*B. subtilis, S. aureus*	*E. coli, P. aeruginosa*	[83]
	P. jaubertii	Leaf extract	Spherical	20–50 nm	*S. aureus*	*E. coli*	[84]
	P. vulgaris	Aerial part extract	Spherical	14–50 nm	*S. aureus*	*E. colil*	[68]

6.5 Nanoparticles as Novel Weapons Against Bacterial Infections

6.5.1 Mode of Action of Bamboo-Derived NPs

Nanoparticles derived from *Bambusa*, specifically AgNPs, exhibit high bactericidal activity by virtue of multiple action mechanisms. They enter the bacterial cell walls, interfere with membrane integrity, and induce ROS production, which degrades proteins, lipids, and nucleic acids. Such nanoparticles interfere with metabolic processes as well as biofilm formation and are effective against both Gram-positive and Gram-negative bacteria like *S. aureus*, *B. subtilis*, and *E. coli* [86]. The fact that they can resist biofilm-mediated resistance highlights their therapeutic significance.

6.5.2 Antibacterial Properties of Amaranthus-Mediated NPs

Amaranthus-mediated nanomaterials function through complementary mechanisms. ZnO NPs damage membrane integrity, dehydrogenase/reductase activity, protein expression, and biofilm development in *P. aeruginosa* and *B. subtilis* [87]. Ag-ZnO nanocomposites further destroy *P. aeruginosa* biofilms that have been allowed to mature and inhibit new formation through ROS and membrane damage [88]. The plant-capped AgNPs of *A. dubius* and *A. viridis* provide coverage against *E. coli*, *Salmonella* spp., and *S. aureus*, are bactericidal, and have low MICs; the phytochemical capping layer enables stability and surface interactions to enhance killing [79, 81].

6.5.3 Pulicaria-Based Nanoparticles as Multifunctional Antimicrobials

AgNPs of *P. glutinosa* and AgCl-NPs from *P. vulgaris* have demonstrated excellent broad-spectrum antibacterial activity. Their mode of action has been suggested through multiple pathways, including cell membrane disruption, induction of oxidative stress, inhibition of some key enzymes, and interference with bacterial DNA [68, 89]. Phytochemical capping enhances their performance by improving their solubility and stability, which increases the strength of bacterial surface binding and, at times, enhances the effects of other antibiotics, such as amoxicillin, a factor that helps attain low MICs even against resistant bacterial strains. With their ability to act on membranes, biofilms, enzymes, and DNA at the same time, *Pulicaria*-based nanoparticles stand out as a powerful strategy against multidrug-resistant bacteria.

6.6 Challenges and Future Directions

Despite the promise of PDNPs, there are some challenges to overcome, including non-standard methods of synthesis that have contributed to inconsistency in size and drug potential, as well as typically sparse in vivo studies to know actual toxicity [90]. Additionally, there are also environmental concerns surrounding the potential for accumulation or persistence of nanoparticles in the environment, meaning greener protocols of both synthesis and disposal will be critical [91]. The standardization of protocols, diligent biosafety testing, and more specific, harmonized regulations are also central to informing the common use of PDNPs as a therapeutic option in the future.

6.7 Conclusion

The rise in antimicrobial resistance has made conventional antibiotics less effective; there is an immediate need for innovative alternative strategies. These nanoparticles include Ag, small nanoparticles, 20–80 nm, spherical, and made from underutilized plants such as *Bambusa* and other plants. These may represent renewable, biodegradable, and powerful antimicrobial agents. These green synthesized nanoparticles have a higher phytochemical content, thus offering a potential for doing multimodal action against Gram-positive organisms, like *B. subtilis* and *S. aureus*, as well as Gram-negative organisms, like *E. coli* and *P. aeruginosa*, which may also include resistant strains. Although at a disadvantage due to its requirements for safety validation, standardization, and harvesting in bulk, more cross-disciplinary studies in the near future may guide the establishment of these nanoparticles as sustainable next-generation therapeutics and prophylactics for a host of global health issues, combining ancient knowledge of plants and contemporary nanotechnology.

Author's Contribution IB, ST—Writing original draft, data collection; IB, ST, NB—Data interpretation, conceptualization; KCK, HM, NB—Final formatting, final approval; IB, ST—Data collection.

Permission to Reproduce from Other Sources None.

Ethical Approval Statement None.

Conflicts of Interest Authors declare no conflicts of interest.

References

1. Oliveira M, Antunes W, Mota S et al (2024) An overview of the recent advances in antimicrobial resistance. Microorganisms 12(9):1920. https://doi.org/10.3390/microorganisms12091920
2. Aijaz M, Ahmad M, Ansari MA et al (2023) Antimicrobial resistance in a globalized world: current challenges and future perspectives. Int J Pharm Drug Des 1(1):7–22. https://doi.org/10.62896/1wkwt966
3. Gauba A, Rahman KM (2023) Evaluation of antibiotic resistance mechanisms in gram-negative bacteria. Antibiotics 12(11):1590. https://doi.org/10.3390/antibiotics12111590
4. Li G, Walker MJ, De Oliveira DM (2022) Vancomycin resistance in *Enterococcus* and *Staphylococcus aureus*. Microorganisms 11(1):24. https://doi.org/10.3390/microorganisms11010024
5. Hetta HF, Ramadan YN, Al-Harbi AI et al (2023) Nanotechnology as a promising approach to combat multidrug resistant bacteria: a comprehensive review and future perspectives. Biomedicine 11(2):413. https://doi.org/10.3390/biomedicines11020413
6. Parvin N, Joo SW, Mandal TK (2025) Nanomaterial-based strategies to combat antibiotic resistance: mechanisms and applications. Antibiotics 14(2):207. https://doi.org/10.3390/antibiotics14020207
7. Odeku OA, Ogunniyi QA, Ogbole OO et al (2024) Forgotten gems: exploring the untapped benefits of underutilized legumes in agriculture, nutrition, and environmental sustainability. Plants 13(9):1208. https://doi.org/10.3390/plants13091208
8. Shrinet K, Singh RK, Chaurasia AK et al (2021) Bioactive compounds and their future therapeutic applications. In: Natural bioactive compounds. Academic, pp 337–362. https://doi.org/10.1016/B978-0-12-820655-3.00017-3
9. Akhtar J, Patowary L (2022) *Bambusa vulgaris*: a comprehensive review of its traditional uses, phytochemicals and pharmacological activities. Sci Phytochem 1(2):68–75. https://doi.org/10.58920/sciphy01020011
10. Jimoh MO, Afolayan AJ, Lewu FB (2019) Therapeutic uses of *Amaranthus caudatus* L. Trop Biomed 36(4):1038–1053
11. Castagliuolo G, Dell'Annunziata F, Pio S et al (2024) Spectroscopic characterization and biological effects of 1-oxo-bisabolone-rich *Pulicaria burchardii* Hutch. subsp. burchardii essential oil against viruses, bacteria, and spore germination. Plants 14(1):68. https://doi.org/10.3390/plants14010068
12. Anju T, Kumar A (2024) Traditional ecological knowledge and medicinal plant diversity usage among the Mullu Kuruman tribes of Wayanad district of Kerala, India and its implications for biodiversity conservation in the face of climate change. Trees For People 16:100595. https://doi.org/10.1016/j.tfp.2024.100595
13. Mohammed HA, Al-Omar MS, Khan RA et al (2021) Chemical profile, antioxidant, antimicrobial, and anticancer activities of the water–ethanol extract of *Pulicaria undulata* growing in the oasis of central Saudi Arabian desert. Plants 10(9):1811. https://doi.org/10.3390/plants10091811
14. Karnwal A, Jassim AY, Mohammed AA et al (2024) Nanotechnology for healthcare: plant-derived nanoparticles in disease treatment and regenerative medicine. Pharmaceuticals 17(12):1711. https://doi.org/10.3390/ph17121711
15. Fatima M, Aqib AI, Faraz H et al (2025) Neutering pathogens through green synthesized nanoparticles. Microb Pathog:107495. https://doi.org/10.1016/j.micpath.2025.107495
16. Franci G, Falanga A, Galdiero S et al (2015) Silver nanoparticles as potential antibacterial agents. Molecules 20(5):8856–8874. https://doi.org/10.3390/molecules20058856
17. More PR, Pandit S, Filippis AD et al (2023) Silver nanoparticles: bactericidal and mechanistic approach against drug resistant pathogens. Microorganisms 11(2):369. https://doi.org/10.3390/microorganisms11020369

18. Khurana SK, Tiwari R, Sharun K et al (2019) *Emblica officinalis* (Amla) with a particular focus on its antimicrobial potentials: a review. J Pure Appl Microbiol 13(4):1995–2012. https://doi.org/10.22207/JPAM.13.4.11
19. Diksha D, Pai MO, Gupta P et al (2023) Investigating the antibacterial, anti-motility and anti-biofilm potential of *Syzygium cumini* mediated silver nanoparticles against emerging multi-drug-resistant nosocomial pathogens. Preprint available at Research Square. https://doi.org/10.21203/rs.3.rs-2904921/v1
20. Sarker U, Rabbani MG, Oba S et al (2022) Phytonutrients, colorant pigments, phytochemicals, and antioxidant potential of orphan leafy *Amaranthus* species. Molecules 27(9):2899. https://doi.org/10.3390/molecules27092899
21. Rahman A, Rauf A, Ali B et al (2022) Phytochemical analysis and antibacterial activity of *Berberis vulgaris* extract. Adv Life Sci 9(3):289–294. https://doi.org/10.62940/als.v9i3.1222
22. Pulipati S, Srinivasa Babu P, Lakshmi NM (2015) Phytochemical analysis and antibacterial efficacy of *Amaranthus* tricolor (L.) methanolic leaf extract against clinical isolates of urinary tract pathogens. Afr J Microbiol Res 9(20):1381–1385. https://doi.org/10.5897/AJMR2015.7294
23. Ugbogu OC, Emmanuel O, Agi GO et al (2021) A review on the traditional uses, phytochemistry, and pharmacological activities of clove basil (*Ocimum gratissimum* L.). Heliyon 7(11). https://doi.org/10.1016/j.heliyon.2021.e08404
24. Huang W, Wang Y, Tian W et al (2022) Biosynthesis investigations of terpenoid, alkaloid, and flavonoid antimicrobial agents derived from medicinal plants. Antibiotics 11(10):1380. https://doi.org/10.3390/antibiotics11101380
25. Wiles D, Pearson JS, Beddoe T (2025) Harnessing plant-derived terpenoids for novel approaches in combating bacterial and parasite infections in veterinary and agricultural settings. Curr Microbiol 82(4):134. https://doi.org/10.1007/s00284-025-04113-4
26. Jadimurthy R, Jagadish S, Nayak SC et al (2023) Phytochemicals as invaluable sources of potent antimicrobial agents to combat antibiotic resistance. Life 13(4):948. https://doi.org/10.3390/life13040948
27. Dong S, Yang X, Zhao L et al (2020) Antibacterial activity and mechanism of action saponins from *Chenopodium quinoa* Willd. husks against foodborne pathogenic bacteria. Ind Crops Prod 149:112350. https://doi.org/10.1016/j.indcrop.2020.112350
28. Kalyan N, Indira A, Joshi B et al (2024) An overview of antimicrobial potential of bamboo. In: Proceedings of the 12th World Bamboo Congress, Taiwan, 18–22 April 2024
29. Schröter D, Baldermann S, Schreiner M et al (2018) Natural diversity of hydroxycinnamic acid derivatives, flavonoid glycosides, carotenoids and chlorophylls in leaves of six different *Amaranthus* species. Food Chem 267:376–386. https://doi.org/10.1016/j.foodchem.2017.11.043
30. Alamdary MP, Baharfar R (2023) Phytochemicals and biological activities of *Pulicaria* genus: emphasis on the flavonoids and sesquiterpenoids and cytotoxicity effects. Curr Org Chem 27(6):526–539. https://doi.org/10.2174/1385272827666230522152708
31. Al-Ghanayem AA, Alhussaini MS, Asad M et al (2022) Effect of *Moringa oleifera* leaf extract on excision wound infections in rats: antioxidant, antimicrobial, and gene expression analysis. Molecules 27(14):4481. https://doi.org/10.3390/molecules27144481
32. Chiş A, Noubissi PA, Pop OL et al (2023) Bioactive compounds in *Moringa oleifera*: mechanisms of action, focus on their anti-inflammatory properties. Plants 13(1):20. https://doi.org/10.3390/plants13010020
33. Alzohairy MA (2016) Therapeutics role of *Azadirachta indica* (Neem) and their active constituents in diseases prevention and treatment. Evid Based Complement Alternat Med 2016:7382506. https://doi.org/10.1155/2016/7382506
34. Herrera-Calderon O, Ejaz K, Wajid M et al (2019) *Azadirachta indica*: antibacterial activity of neem against different strains of bacteria and their active constituents as preventive in various diseases. Pharm J 11(6s):1597–1604. https://doi.org/10.5530/pj.2019.11.244

35. Gupta A, Gupta P, Bajpai G (2024) Tinospora cordifolia (Giloy): an insight on the multifarious pharmacological paradigms of a most promising medicinal ayurvedic herb. Heliyon 10(4):e26125. https://doi.org/10.1016/j.heliyon.2024.e26125
36. Imran M, Khan S (2017) Antibacterial activity of *Syzigium cumini* leaf extracts against multidrug resistant pathogenic bacteria. J Appl Pharm Sci 7(3):168–174. https://doi.org/10.7324/JAPS.2017.70327
37. Prabhu KS, Lobo R, Shirwaikar AA et al (2009) *Ocimum gratissimum*: a review of its chemical, pharmacological and ethnomedicinal properties. Open Complement Med J 1(1):1–15. https://doi.org/10.2174/1876391X00901010001
38. Xie Y, Zhang C, Mei J et al (2023) Antimicrobial effect of *Ocimum gratissimum* L. essential oil on *Shewanella putrefaciens*: insights based on the cell membrane and external structure. Int J Mol Sci 24(13):11066. https://doi.org/10.3390/ijms241311066
39. Zhakipbekov K, Turgumbayeva A, Akhelova S et al (2024) Antimicrobial and other pharmacological properties of *Ocimum basilicum* (Lamiaceae). Molecules 29(2):388. https://doi.org/10.3390/molecules29020388
40. Reyes DC, Ma Z, Romero JJ (2024) The antimicrobial properties of technical lignins and their derivatives—a review. Polymers 16(15):2181. https://doi.org/10.3390/polym16152181
41. Waditzer M, Bucar F (2021) Flavonoids as inhibitors of bacterial efflux pumps. Molecules 26(22):6904. https://doi.org/10.3390/molecules26226904
42. Piplani P, Kumar A, Kulshreshtha A et al (2024) Recent development of DNA gyrase inhibitors: an update. Mini Rev Med Chem 24(10):1001–1030. https://doi.org/10.2174/0113895575264264230921080718
43. Xie Y, Chen J, Wang B et al (2022) Inhibition of quorum-sensing regulator from *Pseudomonas aeruginosa* using a flavone derivative. Molecules 27(8):2439. https://doi.org/10.3390/molecules27082439
44. Sun W, Shahrajabian MH (2023) Therapeutic potential of phenolic compounds in medicinal plants—natural health products for human health. Molecules 28(4):1845. https://doi.org/10.3390/molecules28041845
45. Angelini P (2024) Plant-derived antimicrobials and their crucial role in combating antimicrobial resistance. Antibiotics 13(8):746. https://doi.org/10.3390/antibiotics13080746
46. Kaundal R, Kumar D (2025) Current demands for standardization of Indian medicinal plants: a critical review. Med Drug Discov:100211. https://doi.org/10.1016/j.medidd.2025.100211
47. Arip M, Selvaraja M, R M (2022) Review on plant-based management in combating antimicrobial resistance–mechanistic perspective. Front Pharmacol 13:879495. https://doi.org/10.3389/fphar.2022.879495
48. Allemailem KS (2021) Antimicrobial potential of naturally occurring bioactive secondary metabolites. J Pharm Bioallied Sci 13(2):155–162. https://doi.org/10.4103/jpbs.JPBS_753_20
49. Khan F, Shariq M, Asif M et al (2022) Green nanotechnology: plant-mediated nanoparticle synthesis and application. Nanomaterials 12(4):673. https://doi.org/10.3390/nano12040673
50. Chakraborty N, Jha D, Roy I et al (2022) Nanobiotics against antimicrobial resistance: harnessing the power of nanoscale materials and technologies. J Nanobiotechnol 20(1):375. https://doi.org/10.1186/s12951-022-01573-9
51. Gupta C, Gupta MK, Tripathi S (2025) Nanotechnology: a promising technology in sustainable agriculture. Discov Biotechnol 2(1):12. https://doi.org/10.1007/s44340-025-00022-1
52. Teow SY, Wong MMT, Yap HY et al (2018) Bactericidal properties of plant-derived metal and metal oxide nanoparticles (NPs). Molecules 23(6):1366. https://doi.org/10.3390/molecules23061366
53. Rotello VM (2023) Nanomaterials for fighting multidrug-resistant biofilm infections. BME Front 4:0017. https://doi.org/10.34133/bmef.0017
54. Zhang S, Lin L, Huang X et al (2022) Antimicrobial properties of metal nanoparticles and their oxide materials and their applications in oral biology. J Nanomater 2022(1):2063265. https://doi.org/10.1155/2022/2063265

55. Bhardwaj I, Kumar V, Bhardwaj N et al (2023) A glimpse into the performance and synthesis of microbial nanoparticles and its new advances in soil enrichment and plant nutrition: a review. Nanotechnol Environ Eng 8(4):943–964. https://doi.org/10.1007/s41204-023-00336-7
56. Balachandran A, Sreenilayam SP, Madanan K et al (2022) Nanoparticle production via laser ablation synthesis in solution method and printed electronic application—a brief review. Results Eng 16:100646. https://doi.org/10.1016/j.rineng.2022.100646
57. Hano C, Abbasi BH (2021) Plant-based green synthesis of nanoparticles: production, characterization and applications. Biomolecules 12(1):31. https://doi.org/10.3390/biom12010031
58. Eker F, Akdaşçi E, Duman H et al (2025) Green synthesis of silver nanoparticles using plant extracts: a comprehensive review of physicochemical properties and multifunctional applications. Int J Mol Sci 26(13):6222. https://doi.org/10.3390/ijms26136222
59. Thakur S, Gupta R (2025) Medicinal plants derived green synthesis of nanoparticles: classification, methods, applications and environmental impacts. Nanotechnol Environ Eng 10(2):27. https://doi.org/10.1007/s41204-025-00421-z
60. Khalifa HO, Alkhoori H (2025) Beyond the glitter: gold nanoparticles as powerful weapons against multi-drug resistant pathogens. Front Mol Biosci 12:1612526. https://doi.org/10.3389/fmolb.2025.1612526
61. Macêdo HLRDQ, de Oliveira LL, de Oliveira DN et al (2024) Nanostructures for delivery of flavonoids with antibacterial potential against Klebsiella pneumoniae. Antibiotics 13(9):844. https://doi.org/10.3390/antibiotics13090844
62. Mosa KA, El-Naggar M, Ramamoorthy K et al (2018) Copper nanoparticles induced genotoxicity, oxidative stress, and changes in superoxide dismutase (SOD) gene expression in cucumber (*Cucumis sativus*) plants. Front Plant Sci 9:872. https://doi.org/10.3389/fpls.2018.00872
63. Girma A, Mebratie G, Mekuye B et al (2024) Antibacterial capabilities of metallic nanoparticles and influencing factors. Nano Select 5(12):e202400049. https://doi.org/10.1002/nano.202400049
64. Khalif OI, Alkalifawi EJ (2025) Green-synthesized zinc oxide nanoparticles for *Acinetobacter baumannii* control: a review of plant-based approaches. Cureus 17(5). https://doi.org/10.7759/cureus.83506
65. Enggardipta RA, Akizuki M, Sekine K et al (2025) Antibacterial efficacy of chitosan nanoparticles against *Enterococcus faecalis* in planktonic and biofilm forms. J Appl Microbiol 136(7):lxaf174. https://doi.org/10.1093/jambio/lxaf174
66. Rukavina Z, Vanić Ž (2016) Current trends in development of liposomes for targeting bacterial biofilms. Pharmaceutics 8(2):18. https://doi.org/10.3390/pharmaceutics8020018
67. Salayová A, Bedlovičová Z, Daneu N et al (2021) Green synthesis of silver nanoparticles with antibacterial activity using various medicinal plant extracts: morphology and antibacterial efficacy. Nanomaterials 11(4):1005. https://doi.org/10.3390/nano11041005
68. Sharifi-Rad M, Pohl P (2020) Synthesis of biogenic silver nanoparticles (AgCl-NPs) using a *Pulicaria vulgaris* Gaertn. aerial part extract and their application as antibacterial, antifungal and antioxidant agents. Nanomaterials 10(4):638. https://doi.org/10.3390/nano10040638
69. Mulat M, Banicod RJS, Tabassum N et al (2025) Multiple strategies for the application of medicinal plant-derived bioactive compounds in controlling microbial biofilm and virulence properties. Antibiotics 14(6):555. https://doi.org/10.3390/antibiotics14060555
70. Alorfi H, Alomari M, Bawakid N et al (2025) Green synthesis of AgNPs using *Forsskaolea tenacissima*: sustainable nanotechnology for antimicrobial, antioxidant, and catalytic activities. Front Nanotechnol 7:1587084. https://doi.org/10.3389/fnano.2025.1587084
71. Azad A, Zafar H, Raza F et al (2023) Factors influencing the green synthesis of metallic nanoparticles using plant extracts: a comprehensive review. Pharm Fronts 5(3):e117–e131. https://doi.org/10.1055/s-0043-1774289
72. Ghosh S, Sarkar T, Chakraborty R (2023) Underutilized plant sources: a hidden treasure of natural colors. Food Biosci 52:102361. https://doi.org/10.1016/j.fbio.2023.102361
73. Lithi IJ, Nakib KIA, Chowdhury AS et al (2025) A review on the green synthesis of metal (Ag, Cu, and Au) and metal oxide (ZnO, MgO, Co_3O_4, and TiO_2) nanoparticles using plant extracts for developing antimicrobial properties. Nanoscale Adv 7(9):2446–2473. https://doi.org/10.1039/D5NA00037H

74. Prabula SS, Hentry C, Al-Farraj S et al (2024) Activity of *Bambusa vulgaris* extract in reducing silver nanoparticles: evaluation against methylene blue organic pollutant and microbial agents. Discov Appl Sci 6(4):142. https://doi.org/10.1007/s42452-024-05784-7
75. Jayarambabu N, Velupla S, Akshaykranth A et al (2023) Bambusa arundinacea leaves extract-derived AgNPs: evaluation of the photocatalytic, antioxidant, antibacterial, and anticancer activities. Appl Phys A Mater Sci Process 129(1):13. https://doi.org/10.1007/s00339-022-06279-1
76. Hassan WA, Mohammed AE, AlShaye NA et al (2024) Characterization of *Amaranthus* species: ability in nanoparticles fabrication and the antimicrobial activity against human pathogenic bacteria. PeerJ 12:e16708. https://doi.org/10.7717/peerj.16708
77. Sandhiya S, Subhasree N, Shivapriya S et al (2013) In vitro antioxidant and antimicrobial potential of *Bambusa arundinacea* (Retz.) Willd. Int J Pharm Pharm Sci 5(2):359–362
78. Kaur HP, Kaur S, Prasad B et al (2015) Phytochemical, antioxidant and antibacterial studies on *Bambusa arundinacea* and *Mangifera indica*. Int J Pure Appl Biosci 3:87–93
79. Koyyati R, Nagati V, Nalvothula R et al (2014) Antibacterial activity of silver nanoparticles synthesized using *Amaranthus viridis* twig extract. Int J Res Pharm Sci 5(1):32–39
80. Kolya H, Maiti P, Pandey A et al (2015) Green synthesis of silver nanoparticles with antimicrobial and azo dye (Congo red) degradation properties using Amaranthus gangeticus Linn leaf extract. J Anal Sci Technol 6(1):33. https://doi.org/10.1186/s40543-015-0074-1
81. Tiwari A, Shaik AH, Brianna B et al (2025) Anti-cancer and antimicrobial efficacy of ZnO nanoparticles synthesized via green route using *Amaranthus dubius* (Spleen Amaranth) leaves extract. Green Chem Lett Rev 18(1):2453532. https://doi.org/10.1080/17518253.2025.2453532
82. Elias EM, Meganathan G, Pappuswamy M (2025) Green synthesis, characterization, and biological applications of silver nanoparticles from Pachira glabra leaf extract. The Microbe:100413. https://doi.org/10.1016/j.microb.2025.100413
83. El-Amier YA, Abduljabbar BT, El-Zayat MM et al (2023) Synthesis of metal nanoparticles via *Pulicaria undulata* and an evaluation of their antimicrobial, antioxidant, and cytotoxic activities. Chemistry 5(4):2075–2093. https://doi.org/10.3390/chemistry5040141
84. Alshameri AW, Abdualgalil B, Mutahar M (2025) Green biogenic fabrication of AgNPs derived from *Pulicaria jaubertii* leaf extract: investigation of antimicrobial and cytotoxic properties. OpenNano:100248. https://doi.org/10.1016/j.onano.2025.100248
85. Menchavez MT, Catipay JD, Espra AS et al (2018) Antibacterial properties of *Bambusa vulgaris* (Bamboo) leaves and *Eryngium foetidum* (Culantro) leaves against *Staphylococcus aureus* and *Escherichia coli*. J Med Plants Res 6(1):6–13. https://doi.org/10.22587/gjmpr.2018.6.1.2
86. Nirmala C, Bajwa HK, Oinam S (2024) Bamboo-mediated green synthesis of silver nanoparticles: a new approach towards utilization of an underutilized plant. Adv Bamboo Sci 6:100061. https://doi.org/10.1016/j.bamboo.2024.100061
87. Hsueh YH, Ke WJ, Hsieh CT et al (2015) ZnO nanoparticles affect *Bacillus subtilis* cell growth and biofilm formation. PLoS One 10(6):e0128457. https://doi.org/10.1371/journal.pone.0128457
88. Alhosani F, Islayem D, Almansoori S, Zaka A et al (2025) Antibiofilm activity of ZnO–Ag nanoparticles against *Pseudomonas aeruginosa*. Sci Rep 15(1):17321. https://doi.org/10.1038/s41598-025-02372-6
89. Spizzirri UG, Aiello F, Carullo G et al (2021) Nanotechnologies: an innovative tool to release natural extracts with antimicrobial properties. Pharmaceutics 13(2):230. https://doi.org/10.3390/pharmaceutics13020230
90. Dikshit PK, Kumar J, Das AK et al (2021) Green synthesis of metallic nanoparticles: applications and limitations. Catalysts 11(8):902. https://doi.org/10.3390/catal11080902
91. Gupta D, Boora A, Thakur A et al (2023) Green and sustainable synthesis of nanomaterials: recent advancements and limitations. Environ Res 231:116316. https://doi.org/10.1016/j.envres.2023.116316

Chapter 7
Underutilized Plant-Derived Nanoparticles Against Pathogenic Yeast

Gözde Koşarsoy Ağçeli, Lara Akay, and Sıla Hoşoğlu

Abstract Pathogenic yeasts have become a major concern in medical mycology due to their opportunistic nature, increasing prevalence, and rising resistance to conventional antifungal agents. Infections caused by *Candida*, *Cryptococcus*, *Trichosporon*, and *Malassezia* species lead to high morbidity and mortality, especially in immunocompromised individuals. The therapeutic limitations of azoles, polyenes, and echinocandins, together with multidrug-resistant strains such as *Candida auris*, highlight the need for novel antifungal strategies. Nanotechnology offers an innovative solution, particularly through the green synthesis of nanoparticles using plant-derived extracts. Phytochemicals, including polyphenols, flavonoids, terpenoids, and alkaloids, act as reducing, capping, and stabilizing agents, enabling sustainable production of biocompatible nanoparticles. These biogenic nanomaterials exhibit antifungal activity via multiple mechanisms: disrupting cell wall and membrane integrity, generating reactive oxygen species (ROS), and inhibiting quorum sensing and biofilm formation. Silver and iron oxide nanoparticles synthesized from plant sources demonstrate strong antifungal efficacy, while polymer-based or phenolic-coated systems improve stability and controlled release. Despite promising in vitro outcomes, challenges persist regarding synthesis standardization, large-scale production, toxicity, and regulatory approval. This chapter highlights plant-derived nanoparticles as sustainable antifungal candidates against pathogenic yeasts, addressing resistance and advancing future biomedical applications.

Keywords Pathogenic yeasts · Antifungal resistance · Nanotechnology · Green Synthesis · Plant-derived nanoparticles

---

G. Koşarsoy Ağçeli (✉) · L. Akay · S. Hoşoğlu
Faculty of Science, Department of Biology, Hacettepe University, Ankara, Turkey
e-mail: gozdekosarsoy@hacettepe.edu.tr

© The Author(s), under exclusive license to Springer Nature Switzerland AG 2026

E. Nepovimova et al. (eds.), *Underutilized Plants Based-Nanoparticles: Herbal Remedy to Treat Human Diseases*, https://doi.org/10.1007/978-3-032-20384-7_7

Abbreviations

AgNPs	Silver nanoparticles
CuONPs	Copper oxide nanoparticles
EC50	Half-maximal effective concentration
FCC	Face-centered cubic
IONPs	Iron oxide nanoparticles
MIC	Minimum inhibitory concentration
MFC	Minimum fungicidal concentration
NPs	Nanoparticles
OC	Oligo-chitosan
PNPs	Plant-based nanoparticles
QS	Quorum sensing
ROS	Reactive oxygen species
SPR	Surface plasmon resonance
TSCA	Toxic Substances Control Act
XRD	X-ray diffraction
ZnONPs	Zinc oxide nanoparticles

7.1 Introduction

7.1.1 Pathogenic Yeasts and Their Impact on Public Health

Yeasts are ubiquitous unicellular eukaryotic microorganisms that play significant roles in ecological balance and various industrial processes. While the majority of yeast species are considered non-pathogenic and even beneficial in many contexts, certain species have emerged as important opportunistic pathogens, particularly in individuals with compromised immune systems [1]. Opportunistic yeast infections have become a growing concern for public health in recent years due to their increasing prevalence, association with healthcare settings, and the emergence of antifungal-resistant strains [2].

Among the clinically relevant pathogenic yeasts, the genera *Candida*, *Cryptococcus*, *Trichosporon*, and *Malassezia* are of particular importance. These organisms are responsible for a wide spectrum of infections ranging from superficial mucosal involvement to life-threatening systemic diseases [3]. *Candida* species, especially *Candida albicans*, represent the most common cause of opportunistic fungal infections in humans. Other non-albicans *Candida* species, including *C. glabrata*, *C. tropicalis*, *C. parapsilosis*, and *C. krusei*, have also gained prominence in recent decades due to their variable antifungal susceptibility profiles. Candidiasis encompasses a broad clinical spectrum from superficial infections affecting mucosal surfaces such as the oral cavity and vagina to invasive candidiasis involving the bloodstream and deep-seated organs. The incidence of these infections has been notably increasing in immunocompromised individuals, patients in intensive care units, and those subjected to invasive medical procedures or broad-spectrum antibiotic therapies [3].

Cryptococcus neoformans and *Cryptococcus gattii* are encapsulated yeasts capable of causing severe systemic infections, most notably cryptococcal meningitis, which is associated with high morbidity and mortality rates, particularly in patients with advanced human immunodeficiency virus/acquired immunodeficiency syndrome (HIV/AIDS) or other forms of immunosuppression. These pathogens are typically acquired through inhalation of environmental propagules, often associated with pigeon droppings or contaminated soil. Their virulence is largely attributed to their polysaccharide capsule, which enables evasion of host immune responses and facilitates dissemination within the host [4]. *Trichosporon* species, particularly *Trichosporon asahii*, are recognized as emerging pathogens in hematological malignancies and other immunocompromised states, often causing invasive fungemia and disseminated infections [4]. *Malassezia* species, although predominantly part of the normal skin flora, are implicated in superficial skin disorders such as pityriasis versicolor and seborrheic dermatitis. In rare instances, particularly among neonates and critically ill patients, *Malassezia* spp. may cause systemic infections, including catheter-associated fungemia [4].

The pathogenesis of pathogenic yeasts involves multiple factors, including adherence to host surfaces, morphogenetic transitions (e.g., hyphal formation in *Candida* spp.), secretion of hydrolytic enzymes (proteases, phospholipases), and the ability to form resilient biofilms on biotic and abiotic surfaces. Biofilm formation is a critical virulence factor as it confers protection against antifungal agents and host immune defenses, thereby contributing to persistent and recurrent infections [5]. From a public health perspective, yeast infections impose a significant burden not only through direct morbidity and mortality but also through increased healthcare costs associated with prolonged hospitalization and complex treatment regimens. The emergence of multidrug-resistant species, such as *Candida auris*, poses a substantial threat to infection control efforts and underscores the need for continuous surveillance and stewardship programs [5].

Effective management of yeast-related infections necessitates early diagnosis, the application of appropriate antifungal therapies, and rigorous infection control measures, particularly in healthcare settings. Moreover, public health strategies should focus on raising awareness among at-risk populations and healthcare providers regarding risk factors and preventive measures. Ongoing research into novel antifungal agents and alternative therapeutic approaches remains essential to mitigate the growing threat posed by pathogenic yeasts to global public health [5].

7.1.2 Mechanisms of Resistance to Conventional Antifungal Agents

The number of antifungal agents available for the treatment of fungal infections remains limited, with most drugs grouped into three primary classes: azoles, polyenes, and echinocandins. These antifungal agents exert either fungistatic or fungicidal effects by targeting specific cellular components and biosynthetic pathways critical for fungal survival. However, due to the increased and often inappropriate use of antifungal drugs in clinical settings, as well as selective pressures from environmental sources, resistance rates among opportunistic fungal pathogens—particularly *Candida*, *Aspergillus*, and *Cryptococcus* species—have risen significantly in recent years. Although resistance mechanisms can vary between fungal species, they are largely based on common molecular strategies. Genetic mutations, altered gene expression levels of target proteins, overexpression of efflux pumps, biofilm formation, and metabolic adaptations are the principal mechanisms contributing to antifungal resistance [6]. Azole antifungals exert their effects by inhibiting lanosterol 14-α-demethylase (Erg11p, CYP51), a key enzyme in the ergosterol biosynthetic pathway, which is essential for maintaining fungal cell membrane integrity. The primary mechanism of resistance to azoles involves point mutations in the ERG11 gene, which alter the enzyme's conformation, reducing azole binding affinity while preserving enzymatic function. Additionally, overexpression of ERG11 leads to increased enzyme levels within the cell, requiring higher drug concentrations for inhibition [6]. In *C. albicans*, gain-of-function mutations in transcription factors such as UPC2 can enhance ERG11 expression, further contributing to resistance. Another significant resistance mechanism involves the overexpression of efflux pumps, which actively export azole compounds from the cell, reducing their intracellular accumulation [7]. Mutations in other enzymes of the ergosterol biosynthesis pathway, such as ERG3, can mitigate the toxic accumulation of sterol intermediates, allowing fungal cells to survive despite azole treatment. In *C. albicans*, chromosomal rearrangements such as isochromosome 5L [i(5L)] formation result in increased gene dosage of ERG11 and TAC1, further amplifying azole resistance [7]. Echinocandins inhibit β-1,3-D-glucan synthase (Fks1p/Fks2p), a key enzyme responsible for the synthesis of β-1,3-D-glucan, a critical component of the fungal cell wall [6]. Resistance to echinocandins is notably increasing in species such as *Candida glabrata* and *C. auris*. The most well-characterized mechanism involves mutations in hotspot regions of the FKS1 and FKS2 genes, which reduce the binding affinity of echinocandins to the glucan synthase complex. These mutations lead to decreased susceptibility and are directly associated with clinical treatment failures. Additionally, some species compensate for reduced β-glucan synthesis by increasing chitin content in the cell wall, enhancing structural integrity and conferring phenotypic resistance. Activation of the cell wall integrity (CWI) signaling pathway and calcineurin-dependent stress responses further contribute to echinocandin resistance by promoting survival under antifungal stress [7].

The phenomenon of multidrug resistance (MDR), characterized by simultaneous resistance to multiple classes of antifungal agents, has become increasingly prevalent among opportunistic fungal species [8]. *C. auris* is a prime example, frequently exhibiting resistance to azoles, echinocandins, and polyenes concurrently. This resistance results from cumulative genetic alterations affecting MIC and efflux pump genes, as well as epigenetic modifications, chronic activation of stress response pathways, and enhanced biofilm formation. Although cross-resistance between azoles and polyenes is rare, shared alterations in the ergosterol biosynthetic pathway can facilitate such phenomena [8, 9]. The emergence of resistance to conventional antifungal agents not only leads to individual treatment failures but also represents a serious public health concern. Resistant infections are associated with increased mortality rates, prolonged hospital stays, heightened healthcare costs, and more complex infection control protocols. The global dissemination of resistant species, particularly *C. auris*, has escalated into a public health crisis, challenging infection control measures worldwide. Combating antifungal resistance necessitates rational antifungal use, robust surveillance programs, the widespread implementation of molecular diagnostic tools, and the promotion of fundamental research aimed at identifying novel therapeutic targets [9].

7.1.3 Nanotechnology as an Emerging Therapeutic Paradigm

The increasing prevalence of antifungal resistance among pathogenic fungi, coupled with the limited number of effective conventional antifungal agents, has intensified the need for novel therapeutic strategies. Traditional antifungal drugs face significant challenges including toxicity, limited bioavailability, and the rapid emergence of resistant strains. In this context, nanotechnology has emerged as a promising alternative approach with the potential to overcome these limitations and revolutionize antifungal therapy [10]. Nanotechnology refers to the design, synthesis, and application of materials at the nanoscale—typically within the 1–100 nm range—where unique physicochemical properties arise due to high surface area-to-volume ratios and quantum effects. These properties enable nanoparticles (NPs) to interact with biological systems in novel and highly efficient ways, which conventional macroscale materials cannot achieve. Over the past two decades, nanotechnology has expanded across various biomedical fields, including drug delivery, diagnostics, imaging, and regenerative medicine [10].

In antifungal therapy, nanotechnology offers several advantages that can address the critical challenges of current treatment modalities. Nanoparticles can be engineered to improve the solubility, stability, and targeted delivery of antifungal agents, thereby enhancing therapeutic efficacy while reducing systemic toxicity. For instance, encapsulating hydrophobic antifungal drugs such as amphotericin B in liposomal or polymeric nanoparticles improves their pharmacokinetic profiles and decreases nephrotoxicity, a major side effect limiting amphotericin B usage [10].

Another significant advantage of nanotechnology lies in its ability to disrupt fungal biofilms, complex, highly resistant microbial communities that are a major cause of persistent infections and antifungal failure. Nanoparticles can penetrate biofilms effectively and either deliver antifungal agents directly to fungal cells or destabilize the biofilm matrix, enhancing drug susceptibility [9–11]. Recent advances also include stimuli-responsive nanoparticles, which release their antifungal payloads in response to specific microenvironmental triggers such as pH, temperature, or enzymatic activity characteristic of infection sites. This targeted release minimizes off-target effects and maximizes drug concentration at the infection locus. Additionally, conjugation of nanoparticles with ligands such as antibodies or peptides allows for selective binding to fungal cell surface components, enabling precision targeting and sparing of host cells [10].

Despite these promising developments, the translation of nanotechnology-based antifungal therapies from bench to bedside remains in its early stages. Challenges, including nanoparticle toxicity, large-scale manufacturing, regulatory approval, and long-term safety, require careful consideration and rigorous investigation [11]. In summary, nanotechnology offers a versatile platform that can complement and enhance conventional antifungal strategies, addressing issues of drug resistance, toxicity, and bioavailability. By integrating nanomaterials with antifungal agents and exploiting their unique physicochemical and biological properties, next-generation antifungal therapies with improved efficacy and safety profiles are within reach.

7.1.4 Plant-Mediated Green Synthesis of Nanoparticles: A Sustainable Approach

Nanotechnology is a rapidly evolving and multidisciplinary field that involves the manipulation of matter at the molecular and atomic scales. While noble metals like gold and silver are frequently used for NP synthesis through various chemical and physical methods, these approaches are often not ecologically friendly, necessitating the development of non-toxic, green production methods. In recent years, several biological systems, including bacteria, fungi, and plant extracts, have played a significant role in green synthesis strategies (Fig. 7.1). Among these, plant-based NP green synthesis is considered a benchmark due to its simplicity, cost-efficiency, and the vast diversity of available plant resources. This approach leverages plants as natural "chemical factories," utilizing their rich metabolic machinery to fabricate nanoparticles in an environmentally benign manner [12].

The efficacy of plant-mediated synthesis stems from the diverse array of phytochemicals present in plant tissues, including leaves, stems, roots, and fruits. Compounds such as polyphenols, flavonoids, terpenoids, alkaloids, saponins, polysaccharides, and proteins are pivotal to this process, serving concurrently as reducing, capping, and stabilizing agents. The core mechanism involves the reduction of metal ions from a precursor salt to their zerovalent state. For example, bioactive compounds with antioxidant properties, such as flavonoids and polyphenols containing hydroxyl and ketone functional groups, donate electrons to reduce gold cations (Au^{+} or Au^{3+} to elemental gold (Au^{0})). Following this reduction, these same phytochemicals adsorb onto the surface of the newly formed nanoparticles. This organic layer, or corona, prevents the nanoparticles from aggregating, thereby ensuring their colloidal stability in solution. The synthesis process is typically straightforward, often involving the boiling of plant material in distilled water to create an aqueous extract, which is then mixed with a metal salt solution. The

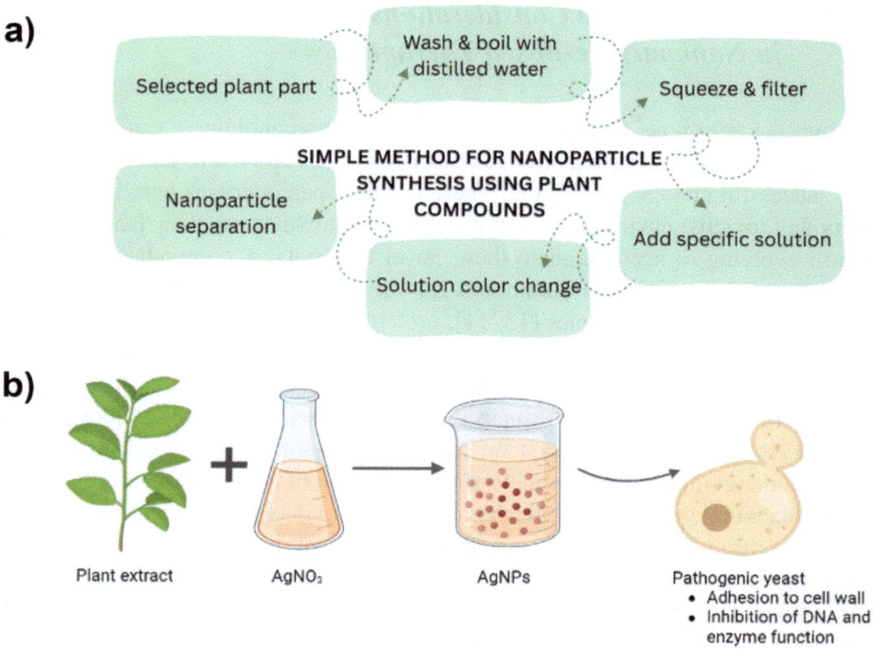

Fig. 7.1 (**a**) A simple method for nanoparticle synthesis using plant compounds, (**b**) synthesis of plant-mediated silver nanoparticles

subsequent change in the solution's color indicates the formation of nanoparticles [13].

This green synthesis route offers substantial advantages over conventional methods. It is inherently sustainable and cost-effective, utilizing inexpensive and readily available raw materials, which can even include agricultural waste, thereby lowering production costs and reducing environmental impact. The process avoids the use of toxic chemicals, high pressures, and extreme temperatures, which eliminates the generation of hazardous byproducts and enhances the biocompatibility of the final product. Furthermore, plant-assisted synthesis often exhibits faster reaction kinetics compared to other biological methods, such as those using bacteria or fungi. An often-understated but critical advantage of this method is the inherent biofunctionalization of the nanoparticles. The phytochemicals that form the stabilizing corona are not merely inert coatings; many possess their own intrinsic antimicrobial, antioxidant, and anti-inflammatory properties. Consequently, the resulting nanoparticle is a hybrid therapeutic agent. It combines the potent, multi-modal antimicrobial actions of the metallic core with the bioactive properties of the plant-derived surface coating, creating a synergistic effect that is likely more powerful than either component in isolation. This natural synergy is a key distinction and a significant advantage of green synthesis over chemical methods that rely on passive, non-bioactive stabilizers [14].

7.1.5 Challenges and Considerations in Nanoparticle-Based Therapeutics

Despite the enormous therapeutic potential of nanotechnology, a critical and fair assessment identifies important issues that need to be resolved to guarantee its safe and successful clinical application. These hurdles span concerns over biocompatibility and toxicity, environmental impact, and a complex regulatory landscape that is still evolving to accommodate these novel materials. A comprehensive understanding of these issues is paramount for the responsible development of next-generation antifungal therapies [13, 14].

The very physicochemical properties that make nanoparticles effective therapeutic agents—their minute size (1–100 nm) and exceptionally high surface area-to-volume ratio—are also the primary drivers of their potential toxicity. When introduced into a biological system, these properties facilitate novel interactions with cellular components that can lead to a spectrum of adverse effects collectively termed nanotoxicity. A central mechanism of this toxicity is the induction of oxidative stress through the excessive generation of reactive oxygen species (ROS). This surge in ROS can overwhelm the cell's natural antioxidant defenses, leading to widespread damage to critical biomolecules, including lipid peroxidation, protein carbonylation, and DNA strand breaks. This can trigger inflammatory responses, mitochondrial dysfunction, and ultimately, programmed cell death or apoptosis. The biodistribution and toxic potential of NPs are highly dependent on their specific characteristics, such as size, shape, surface charge, and chemical composition. Smaller nanoparticles, for instance, are more readily taken up by cells and can traverse critical biological barriers like the blood-brain barrier and the placental barrier, leading to accumulation in sensitive organs such as the brain, liver, kidneys, and spleen, where their long-term health consequences remain largely unknown [9, 12–14].

Beyond direct human health concerns, the lifecycle of nanoparticles from production to disposal raises significant environmental and regulatory questions. The release of engineered nanomaterials into air, water, and soil could have unforeseen and potentially disruptive ecotoxicological effects on various organisms and ecosystem functions. The same unique properties that are beneficial in medicine may render them hazardous under certain environmental conditions. This has created a formidable challenge for regulatory agencies like the US Environmental Protection Agency (EPA). Existing frameworks, such as the Toxic Substances Control Act (TSCA), may not be fully equipped to assess the risks of nanomaterials, which can exhibit profoundly different behaviors compared to their bulk chemical counterparts, even when sharing the same molecular identity. Consequently, the pathway for nanomedicines from laboratory research to clinical approval is exceptionally rigorous, demanding reproducible large-scale manufacturing processes, extensive preclinical safety evaluations, and the establishment of new, standardized regulatory guidelines tailored to the unique attributes of these materials. This reveals a central paradox in nanotherapeutics: the mechanisms that confer high efficacy are intrinsically linked to the mechanisms that pose potential toxicity [15]. The scientific challenge, therefore, is not simply to eliminate these toxic properties but to precisely engineer nanoparticles and therapeutic strategies that maximize toxicity toward the target pathogen while minimizing harm to the host and the environment. This reframes the development process as a complex, multi-parameter optimization problem aimed at widening the therapeutic window for these powerful agents [16].

7.1.6 Underutilized Plants as Novel Bio-factories for Antifungal Nanoparticles

In the quest for sustainable, cost-effective, and novel sources for the green synthesis of nanoparticles, scientific attention is increasingly shifting towards underutilized plant species. This broad category encompasses a diverse range of flora, including neglected and orphan crops, endemic species with limited geographical distribution, widespread invasive weeds, and agricultural or industrial plant waste. These plants represent a vast, underexplored reservoir of phytochemical diversity, offering a rich platform for the next generation of nanobiotechnology [17].

The strategic integration of underutilized plants into the nanoparticle synthesis pipeline presents a compelling, multi-faceted value proposition. From an economic and environmental standpoint, it provides a low-cost and sustainable supply of raw materials. This approach can transform agricultural waste or ecologically problematic invasive species into valuable feedstocks for biomedical applications, aligning with the principles of a circular economy and potentially creating new revenue streams. For instance, finding a high-value use for an invasive weed like *Lantana camara* can create an economic incentive for its management and control, thereby converting an ecological liability into a therapeutic asset [18]. This strategy also opens new frontiers for bioprospecting, as these plants may harbor unique phytochemicals not found in common crops. These novel compounds could confer enhanced bioactivity or unique physicochemical properties to the resulting nanoparticles, such as improved stability, smaller size, or specific surface functionalities. The immense biodiversity within this group of plants offers a virtually limitless library of natural reducing and stabilizing agents, which could allow for the fine-tuning of nanoparticle characteristics for specific applications [18, 19].

This approach establishes a symbiotic relationship between advanced materials science and broader ecological and conservation goals. By demonstrating the unique therapeutic potential of nanoparticles derived from endemic or locally significant plants, such as *Alkanna tinctoria*, a strong case can be made for their conservation and sustainable cultivation. Such research highlights the tangible value of biodiversity and provides a powerful argument against its loss. Therefore, the use of underutilized plants transcends a simple search for convenient synthesis reagents. It represents a holistic strategy that can simultaneously advance nanomedicine, promote sustainable agriculture, contribute to environmental management, and underscore the critical importance of preserving global botanical diversity [20].

7.1.7 Antimicrobial Mechanisms of Plant-Derived Nanoparticles

Nanoparticles (NPs) are extensively utilized to target bacteria as an alternative due to their low-risk side effects [19]. Moreover, recently developed metal-based nanoparticles exhibit antibacterial properties and could be used as new materials to prevent the spread of antimicrobial resistance [21]. Direct contact and subsequent cell damage represent one of the major antibacterial mechanisms. This process involves interactions with the bacterial cell walls and membranes, often mediated through metabolic pathways. Upon contact, plant-derived metallic nanoparticles can cause damaging interactions with cell walls. The damaging interactions may lead to the leakage of intracellular components. These interactions alter redox homeostasis and oxidative stress, leading to the disruption of cell walls; the resulting homeostatic imbalance and loss of membrane integrity allow nanoparticles to enter the bacterial cells.

7.2 Underutilized Plants-Mediated Metal-Based Nanoparticles

7.2.1 Silver Nanoparticles (AgNPs)

AgNPs are known to have promising potential in a variety of fields such as nanotechnology and the biomedical field, due to their unique properties, especially their excellent antibacterial, antifungal, and antiviral effects [22] (Fig. 7.1b). The emerging role of multidrug resistance among pathogenic bacteria and yeast has become a serious medical problem which creates a rationale for AgNPs to be used in combination with antibiotics and fungicides [23]. Although AgNPs can be produced by physical and chemical methods, their biosynthesis mediated by plants (referred to as the green synthesis) is gaining importance owing to its eco-friendliness, cost-effectiveness, and production of fewer toxic by-products. This is an approach, generally conducted in a bottom-up process, in which silver precursors, like silver nitrate ($AgNO_3$), are converted into metallic silver (Ag^0) by phytochemicals in the plant extracts [24]. The bioactive moieties, including flavonoids, alkaloids, polyphenols, terpenoids, saponins, carbohydrates, lipids, and proteins, not only reduce the metal ions but also act as capping and stabilizing agents by providing a protective layer upon the nanoparticle surface, thereby preventing agglomeration. A variety of plant species have been used for this reason. For example, the use of invasive or weedy plant species represents a particularly sustainable and high-value approach, transforming ecologically problematic biomass into advanced biomedical materials. A compelling case is the use of *Melia azedarach* (Chinaberry), an invasive tree species rich in terpenoids and other bioactive compounds. Aqueous leaf extracts of *M. azedarach* have been successfully employed as bioreductants for the synthesis of stable AgNPs. This strategy not only aligns with the principles of green chemistry but also introduces an element of ecological valorization, creating a "bioremediation-to-biomedicine" pathway where the control of an invasive species is directly linked to the production of therapeutic agents. *Citrullus colocynthis* (bitter apple), also a cucurbitacin and known for its terpenoids, alkaloids, saponins, and polyphenols, has been utilized to reduce silver ions by methanolic extracts of its seeds and pulp [25].

7.2.2 Characterization of Underutilized Plants-Mediated Silver Nanoparticles

For a complete comprehension of the bio-activities of the plant-extracted AgNPs and their possible antifungal action, profiling of their physicochemical properties is necessary. Several analytical methods are used to study the size, shape, crystal structure, surface charge, and stability of the nanoparticles. A summary of frequently used characterization techniques and typical results for AgNPs obtained from various plant and microbial origins is given in the following Table 7.1 [26].

Table 7.1 Analytical techniques and typical physicochemical characteristics of AgNPs from plant and microbial sources

Technique	Purpose	Representative findings
UV-visible spectroscopy (UV-Vis)	Confirms nanoparticle formation and stability; identifies surface plasmon resonance (SPR)	*Melia azedarach*: 434 nm *Citrullus colocynthis*: 427 nm *Crocus sativus*: 318 nm *Parthenium hysterophorus*: 455 nm
Transmission electron microscopy (TEM)	Morphology, shape, and size	*Melia azedarach*: Spherical; 12–46 nm *Crocus sativus*: Nanorod morphology; average size 35–50 nm *Parthenium hysterophorus*: Spherical and Hexagonal 27 ± 5 nm & 84 ± 2 nm *Ziziphus spina-christi*: Spherical; 13–30 nm
X-ray diffraction (XRD)	Crystalline structure and phase identification	*Melia azedarach*: 2θ peaks at 38.12, 44.23, 64.51, 77.69° confirm FCC structure *Citrullus colocynthis*: 2θ peaks at 38.83, 44.90, 64.63, 77.99° confirm FCC structure *Crocus sativus*: Bragg's peaks confirm crystalline Fe_2O_3 structure
Zeta potential	Surface charge and colloidal stability	*Melia azedarach*: −22.3 mV

Plants as bio-factories for synthesizing silver nanoparticles (AgNPs) have shown outstanding antifungal activity against a wide range of pathogenic yeasts, providing very promising alternatives against multidrug-resistant strains. The role of size, shape, surface charge, and the capping phytochemicals governs the effectiveness of these nanoparticles. Smaller NPs, usually <20 nm, have been suggested to possess greater antifungal efficacy as they have more surface area and interact better with the fungal cell walls and membranes. Several studies have reported species-specific sensitivity among pathogenic yeasts. *C. albicans*, a common opportunistic pathogen, often requires higher nanoparticle concentrations to achieve significant inhibition, reflecting its relative resistance. In contrast, *Candida lipolytica* and *Candida parapsilosis* are more susceptible, showing pronounced growth suppression at lower AgNP concentrations. Notably, *C. neoformans* demonstrates high sensitivity even at minimal doses, with plant-derived AgNPs substantially reducing cell viability. AgNPs synthesized from *C. colocynthis*, for example, have shown significant inhibitory effects against resistant strains, highlighting the potential of plant-mediated nanoparticles as effective antifungal agents. The antifungal activity of AgNPs is often superior to conventional drugs; for instance, AgNPs have demonstrated greater efficacy against a panel of clinical *Candida* isolates, including *C. albicans*, *C. krusei*, *C. glabrata*, and *C. tropicalis*, than the standard antifungal amphotericin B [27].

7.2.3 Iron Oxide Nanoparticles (IONPs)

Iron is a key component in both the biology and geology of Earth. Its oxides have been widely used as magnetic components, e.g., magnetite (Fe_3O_4) and hematite (Fe_2O_3) in nanotechnology. (hereafter, IONPs) are attractive ones due to their superparamagnetic properties, low toxicity, and incredible tunability, and therefore, these are the most researched ones among metallic nanoparticles [28].

Plant extracts have recently emerged as efficient reagents for IONP synthesis. A particularly innovative approach involves the use of agricultural waste streams, such as the corms of *Crocus sativus* (saffron), which are discarded in large volumes during saffron production. The selection of such materials is not arbitrary; plant storage organs like corms are biologically programmed to be rich reservoirs of carbohydrates, proteins, and a diverse array of secondary metabolites that serve as defense compounds. These are the very same classes of biomolecules that function as potent reducing and capping agents in nanoparticle synthesis. This principle—that the biological function of a plant part can predict its utility in green synthesis—provides a rational basis for selecting new source materials, moving beyond random screening to a more hypothesis-driven approach. Using an aqueous extract of *C. sativus* corms, researchers have successfully synthesized crystalline, nanorod-shaped Fe_2O_3 nanoparticles (Fe_2O_3-CsNPs) with an average size of 35–50 nm. In addition to agricultural byproducts, common but under-investigated plants like *Laurus nobilis* (bay laurel) have been used to produce spherical IONPs with sizes ranging from 8 nm to

46 nm, demonstrating broad-spectrum antifungal activity against phytopathogens like *Alternaria* and *Aspergillus* species. Overall, Fe_2O_3 and Fe_3O_4 NPs have been successfully synthesized from a range of underutilized species to give similar NPs. Usually, the particles do not exceed 20–50 nm and are spherical or cubical in shape, which is dependent on the phytochemicals of the plant source [23–29].

In a lot of research works, it was found that nanoparticles synthesized by these green routes exhibit significant antifungal activity. This is evident in the clear zones of inhibition when Fe_2O_3 nanoparticles were obtained from *Khaya senegalensis* leaf extract on *Candida* species, implying its ability to suppress the growth of yeast. The Fe_2O_3-CsNPs from *C. sativus* demonstrated potent, concentration-dependent inhibition of the phytopathogen *Verticillium dahliae*, achieving up to 86.4% growth reduction and causing visible degeneration of fungal hyphae. This activity translates to pathogenic yeasts, as IONPs are known to be highly effective against *Candida* species, including *C. albicans*, *C. tropicalis*, and *C. glabrata. C. albicans* is the most common pathogenic yeast and, as such, has been widely investigated with respect to nanoparticle-assisted antifungal action. It was shown that plant-derived IONPs inhibit both planktonic cell growth and biofilm, an important virulence factor of *C. albicans*. The nanoparticles compromise membrane integrity, trigger oxidative stress, and reduce the ability of yeast cells to adhere and form biofilms, thereby enhancing susceptibility to host defenses and antifungal drugs. Other clinically important yeasts such as *C. glabrata, C. tropicalis*, and *C. krusei* have also been tested. In one study, Fe_2O_3 nanoparticles synthesized using plant extracts demonstrated growth inhibition against *Candida* tropicalis, with inhibition zones up to 12 mm at concentrations of 500 μg/mL. These findings suggest that IONPs are capable of exerting antifungal effects across multiple *Candida* species, which is clinically relevant given the rising incidence of non-albicans infections [30].

7.2.4 Zinc Oxide Nanoparticles (ZnONPs)

Among the various metal oxide nanoparticles, zinc oxide nanoparticles (ZnONPs) have garnered significant attention for antifungal applications due to their high physicochemical stability, novel surface properties, and low toxicity toward human cells. The green synthesis of ZnONPs further enhances their appeal by providing an eco-friendly and cost-effective production route. Following the principle of valorizing problematic biomass, the noxious and globally invasive weed *Parthenium hysterophorus* has been successfully employed as a source of reducing and capping agents for ZnONP synthesis [31].

An important feature of this synthesis method is the ability to control the resulting nanoparticle morphology by simply adjusting the concentration of the plant extract. This control over structure directly influences biological function, a core tenet of nanotechnology. For instance, using a 25% *P. hysterophorus* leaf extract yields predominantly spherical ZnONPs with a size of 27 ± 5 nm, while a 50% extract produces larger, hexagonal nanoparticles with a size of 84 ± 2 nm [32]. This size- and shape-dependent activity was clearly demonstrated against the plant fungal pathogens *Aspergillus flavus* and *Aspergillus niger*, where the smaller, 27 nm spherical nanoparticles exhibited significantly higher antifungal efficacy. The superior performance of smaller nanoparticles is not a coincidental finding but a direct consequence of their physical properties. Nanoparticle efficacy, particularly when mediated by surface-dependent phenomena like reactive oxygen species (ROS) generation or direct membrane contact, is proportional to the surface-area-to-volume ratio. For a given mass of material, smaller particles provide a vastly greater interactive surface, thereby enhancing their biological mechanism of action. This principle is critical for rationally designing nanoparticles with optimized therapeutic effects.

The antifungal activity of ZnONPs extends robustly to pathogenic yeasts. Studies on *C. albicans* have shown a clear dose-dependent inhibitory effect. The half-maximal effective concentration (EC_{50}) for ZnONPs against *C. albicans* has been estimated at 35.59 μg/mL, with a minimum inhibitory concentration (MIC) of 80 μg/mL. This leads to severe morphological alterations, including the pitting of the cell wall, invagination, hyphal damage, and eventual rupture of the cell membrane, resulting in cell death [33].

7.2.5 Copper Oxide Nanoparticles (CuONPs)

Copper oxide nanoparticles (CuONPs) have emerged as a highly promising class of antifungal agents, distinguished by their low cost, high stability, and potent antimicrobial properties. Copper is also an essential micronutrient for plants, contributing to defense protein synthesis, which adds to the appeal of CuONPs in agricultural contexts. Green synthesis routes for CuONPs have been developed using various underutilized plant resources, such as the leaf extract of *Ziziphus spina-christi*, a drought-resistant medicinal tree known for its rich phytochemical profile. This method yields stable, spherical CuONPs with a particle size in the range of 13–30 nm [34].

CuONPs exhibit potent fungicidal activity against pathogenic yeasts. In vitro studies against *C. albicans* have established both a minimum inhibitory concentration (MIC) and a minimum fungicidal concentration (MFC) of 400 μg/mL. However, a particularly significant attribute of CuONPs is their pronounced activity against fungal biofilms. Biofilm-associated infections, common on medical devices like catheters and implants, represent a major clinical challenge due to their extreme resistance to conventional antifungal drugs. The extracellular matrix of the biofilm acts as a physical barrier, protecting the embedded yeast cells from therapeutic agents and host immune responses. Nanoparticles, due to their small size and unique mechanisms of action, are well-suited to overcome this barrier. Green-synthesized CuONPs have demonstrated the ability to eradicate preformed, mature biofilms of multiple clinically relevant species, including *C. albicans, C. glabrata,* and *C. tropicalis*, by 47–66% at a concentration of 100 μg/mL. Furthermore, a trimetallic nanocomposite (CuO/Ag/

ZnO) synthesized using the same *Z. spina-christi* extract also showed strong activity against *C. albicans* and *C. krusei*. This potent anti-biofilm capability positions plant-derived CuONPs not merely as a substitute for existing antifungals but as a specialized therapeutic tool designed to combat the specific, highly problematic, and often recalcitrant nature of biofilm-associated fungal infections. This suggests a targeted pathway for future translational research, such as the development of nanoparticle-coated medical devices to prevent biofilm formation [32–35].

7.3 Mechanisms of Action Against Pathogenic Yeasts

The growing emergence of antifungal resistance presents a serious problem, emphasizing the importance of studying novel antifungal agents, such as metal nanoparticles, as a potential solution. Nanoparticles cause damage to fungi through various mechanisms, including the generation of reactive oxygen species (ROS), effects on cell wall damage, and disruptions in gene and protein regulation [29].

7.3.1 Disruption of Cell Wall and Membrane Integrity

A major mechanism through which nanoparticles mediate antifungal activity against pathogenic yeasts and filamentous fungi is via the disruption of cell wall and membrane integrity. Exposure to nanoparticles triggers various structural and functional alterations in fungal cells. Changes such as surface shrinkage of the cell walls, cell aggregation, the formation of pits and pores, and overall deformation occur. Microscopic observations have revealed that nanoparticles can directly contact and embed within fungal cell walls during adsorption, resulting in the previous morphological changes. Similar to the cell wall, the inner membranes are also subjected to disruptions, such as alterations in the arrangement of organelles and a reduction in cytoplasmic content [30].

7.3.2 Oxidative Stress and Damage to Cellular Components (DNA, Proteins, and Lipids)

A redox imbalance that results from the production and build-up of reactive oxygen species (ROS) during metabolic activities is known as oxidative stress. Enzymes including endothelial nitric oxide synthase, xanthine oxidase, and NADPH oxidases catalyze oxidative reactions that produce ROS, which include oxygen-free radicals like hydrogen peroxide, superoxide, and hydroxyl radicals, as byproducts of molecular oxygen. Excessive ROS production leads to oxidative damage in essential biomolecules, including biomembranes, proteins, nucleic acids, and lipids. Plant-based metal nanoparticles exert this damage, which constitutes the fundamental mechanism underlying their therapeutic effects, including their antimicrobial and specifically antifungal activities. Plant-derived metal nanoparticles induce lipid peroxidation in fungal cells. By promoting lipid/phospholipid membrane oxidation, they result in the disruption of membrane integrity and the inactivation of enzyme functions. The ROS produced by the antifungal activity can directly damage DNA and mitochondria. Free radicals disrupt DNA replication, which inhibits fungal growth. According to the literature, the previously mentioned damages have been shown to be effective against species such as *C. albicans*, *Saccharomyces cerevisiae*, and *C. glabrata* [35, 36].

7.3.3 Inhibition of Quorum Sensing and Biofilm Formation

C. albicans is a major opportunistic human pathogen, and the biofilm it forms increases its resistance to antifungal agents by approximately 1000-fold. Consequently, this contributes to the high mortality rates observed in fungal infections. Current treatments are often inadequate against these pathogens, leading to recurrent infections and adverse effects associated with high drug dosages. A study demonstrated that oligo-chitosan (OC) modified nanoparticles exhibited strong binding to planktonic *C. albicans* cells. This binding, coupled with high ROS production, resulted in potent inhibition of yeast-to-hyphae transition and early-stage biofilm formation. This inhibition is mediated through the disruption of the fungal quorum sensing (QS) mechanism. For dense and adherent mature biofilms, the smallest nanoparticles (~15 nm) showed the fastest and deepest penetration. In conclusion, these nanoparticles suppress both hyphal and biofilm formation by disrupting QS. This powerful strategy can be optimized based on factors such as binding affinity and high ROS production for early-stage inhibition, and deep penetration combined with sufficient ROS generation for the destruction of mature biofilms [32, 34, 35].

7.4 Knowledge Gaps and Research Challenges

Despite promising in vitro antifungal activity, the use of underutilized plants-extract–derived nanoparticles (NPs) against pathogenic yeasts faces major limitations. Intrinsic variability in plant extracts is perhaps the foremost challenge. The concentration and composition of reducing/capping phytochemicals (e.g., phenolics, flavonoids, terpenoids) vary widely with species, plant part, growth stage, season, and geography. This heterogeneity has a direct impact on NP nucleation and growth, leading to discrepancies in size, shape, and stability between batches. In practice, a protocol for one laboratory's methods likely cannot be directly copied elsewhere. For instance, for the identical metal (Ag) and plant extract, various studies reveal distinct NP shapes and antimicrobial efficacy. Additionally, standard extraction and synthesis protocols are needed but remain minimally developed for most underutilized plants. This lack of standardization hinders reproducibility, as studies report different NP concentrations, assay methods (e.g., agar diffusion vs. broth microdilution), and culture conditions, making results challenging to compare or pool [32].

7.4.1 *In Vitro–In Vivo Correlation and Safety*

Nearly all investigations of plant-based NPs against yeasts are conducted in vitro (e.g., agar diffusion, microbroth dilution). There is a large disconnect between these in vitro findings and their clinical relevance. These NPs have not yet been investigated in animal candidiasis models (e.g., *Galleria mellonella*, mice). The ADME and immune interactions of plant NPs are not well characterized. A recent review noted "suboptimal in vivo studies" and "absence of extensive pharmacokinetics" as significant deficiencies [36]. Without animal data, we cannot know if in vitro doses are relevant (i.e., too high, low, toxic, or safe) or how NPs distribute and are cleared in a host. Additionally, plant NPs could elicit immune or inflammatory responses not predicted by cell culture.

Efficacy models: There is no established in vivo model for underutilized-plant NPs versus *Candida*. Bridging this gap will require validated infection models and dosing studies, which have not been reported. *Toxicity and biocompatibility:* Although green NPs are often assumed to be safer than chemically made ones, systematic toxicity data are scarce. A few reports also provide brief descriptions of cytotoxic assays in mammalian cell lines which are limited in number and variation. A more recent review indicates explicitly further studies that are necessary in NP toxicity and biocompatibility. For example, copper- and cobalt-based plant NPs (which are cheaper than Ag) have strong antifungal effects but raise serious biosafety questions [33]. In vivo safety profiles (e.g., organ accumulation, long-term effects) have not been established for any plant-synthesized NPs. *Regulatory requirements:* Due to these uncertainties, no plant-extract NPs have advanced to clinical trials. Agencies (e.g., FDA, EMA) demand thorough preclinical safety information. One review notes that "strict in vivo evidence has been made mandatory by regulatory agencies to approve nanomedicine formulations" [36]. This information does not currently exist for these NPs, which is a major translational barrier.

7.4.2 *Scale-Up Production and Regulatory Hurdles*

Beyond the scientific gaps, practical and regulatory hurdles also exist. Scaling up green NP production without qualitative deterioration remains a challenge. Underutilized plants may lack defined cultivation protocols, affecting their supply chain. Even with mass culture, downstream processing (e.g., filtration, NPs purification, sludge disposal) is not well-studied. To date, no good manufacturing practice (GMP) guidelines exist for biogenic NPs.

From a regulatory standpoint, plant-extract NPs fall between categories: are they novel drugs, botanical supplements, or something else? No specific guidelines exist for these nano-formulations beyond conventional drug or agricultural regulations. As mentioned, agencies require strong toxicology and in vivo efficacy data for approval. Biocompatibility and environmental effects must also be considered. Biosynthesized NPs may have different organ accumulation patterns than chemical NPs, and their long-term ecological fate is largely unknown. Gathering this information is essential for any real-world application of UPL-NPs.

Finally, key unaddressed needs include methodological inconsistency, poor reproducibility and scalability, limited species coverage, little mechanistic understanding, and significant safety and regulatory unknowns. Their conversion to inorganic forms and their uptake or adsorption onto pathogen cell surfaces are pivotal areas for investigation. These challenges must be systematically addressed through an interdisciplinary approach (including materials science, microbiology, pharmacology, and regulatory science) to develop these promising plant-derived nanoparticles as valuable antifungals.

7.5 Conclusion, Future Perspectives, and Biomedical, Agricultural, and Global Health Applications

Plant-based nanoparticles (PNPs) are emerging as a promising antifungal therapeutic in medicine. For instance, green-fabricated nanofiber dressings containing phytochemicals can accelerate chronic wound healing by exerting antimicrobial and anti-inflammatory effects and sustained drug efflux (Fig. 7.1). Similarly, surfaces of catheters and implants, coated with biofabricated metal-plant-polymer composites, are capable of reducing fungal biofilm formation (*Candida*) [35]. These PNP-based dressings/coatings are biocompatible and multitasking, and therefore they will be promising for future wound care and device coating.

Plant-sourced NPs offer novel tools for crop protection and postharvest preservation. For example, ZnO nanoparticles combined with underutilized plant extracts completely inhibited *C. albicans* on stored strawberries and significantly reduced spoilage yeasts and molds. More broadly, engineered metal and metal-oxide PNPs (e.g., Ag, Cu, ZnO) have demonstrated both antifungal and antimycotoxin activity by inhibiting fungal growth and toxin production in grains and produce. Such particles can be applied as seed treatments, foliar sprays, or bioactive films/coatings to intercept toxigenic fungi (e.g., *Aspergillus*, *Fusarium*) and suppress mycotoxin accumulation from farm to table.

Plant-derived NPs inherently suit sustainability and equity objectives. Green production methods of these products use renewable biomass, emitting as little toxic waste as possible. It can significantly reduce environmental impact. Since

PNPs often retain plant source toxins, as well as a low level of environmental impact in decomposition (due to the diversity and abundance of plant materials), they are much less harmful than chemically synthesized NPs. Moreover, using widely available plants (including crops grown in low-resource regions) lowers manufacturing costs and increases accessibility. In this way, PNPs offer a route to equitable healthcare: cutting-edge antifungals that are both affordable and eco-friendly, helping to address fungal infections in underserved populations and in one-health contexts (human, animal, plant health). Overall, plant-derived nanoparticles present a promising, sustainable alternative to conventional nanomaterials. Their low-cost production, reduced ecological footprint, and potential for broad accessibility position them as powerful tools in advancing both global health equity and environmentally responsible therapies.

CRediT Authorship Contribution Statement Gözde Koşarsoy Ağçeli: Visualization, writing—original draft, writing—review and editing; Lara Akay: Visualization, writing—original draft, writing—review and editing. Sıla Hoşoğlu: Visualization, writing—original draft, writing—review and editing.

Declaration of Competing Interests The authors have no conflicts of interest to declare that are relevant to the content of this chapter.

Ethical Approval This study did not involve human participants or vertebrate animals. Therefore, ethical approval was not required. All experimental procedures were performed in compliance with institutional biosafety and research integrity guidelines.

References

1. Kovács M, Pomázi A, Taczman-Brückner A et al (2025) Detection and identification of foodborne yeasts: an overview of the relevant methods and their evolution. Microorganisms 13:981. https://doi.org/10.3390/microorganisms13050981
2. Fisher MC, Alastruey-Izquierdo A, Berman J et al (2022) Tackling the emerging threat of antifungal resistance to human health. Nat Rev Microbiol 20:557–571. https://doi.org/10.1038/s41579-022-00720-1
3. Pruitt HM, Zhu JC, Riley SP et al (2025) The hidden fortress: a comprehensive review of fungal biofilms with emphasis on Cryptococcus neoformans. J Fungi 11:236. https://doi.org/10.3390/jof11030236
4. Hobi S, Cafarchia C, Romano V et al (2022) Malassezia: zoonotic implications, parallels and differences in colonization and disease in humans and animals. J Fungi 8:708. https://doi.org/10.3390/jof8070708
5. Mallick DC, Kaushik N, Goyal L et al (2025) A comprehensive review of Candidemia and invasive candidiasis in adults: focus on the emerging multidrug-resistant fungus Candida auris. Diseases 13:93. https://doi.org/10.3390/diseases13040093
6. Lu X, Zhou J, Ming Y et al (2025) Next-generation antifungal drugs: mechanisms, efficacy, and clinical prospects. Acta Pharm Sin B 15:3852–3887. https://doi.org/10.1016/j.apsb.2025.06.013

7. Song L, Wang S, Zou H et al (2025) Regulation of Ergosterol biosynthesis in pathogenic fungi: opportunities for therapeutic development. Microorganisms 13:862. https://doi.org/10.3390/microorganisms13040862
8. Siqueira AC, Bernardi GA, Arend LNVS et al (2025) Azole resistance and ERG11 mutation in clinical isolates of Candida tropicalis. J Fungi 11:24. https://doi.org/10.3390/jof11010024
9. Lv Q, Yan L, Jiang Y (2016) The synthesis, regulation, and functions of sterols in Candida albicans: well-known but still lots to learn. Virulence 7:649–659. https://doi.org/10.1080/21505594.2016.1188236
10. Lima SL, Colombo AL, De Almeida Junior JN (2019) Fungal cell wall: emerging antifungals and drug resistance. Front Microbiol 10:2573. https://doi.org/10.3389/fmicb.2019.02573
11. Al Musaimi O (2025) FDA-approved antibacterials and echinocandins. Antibiotics 14:166. https://doi.org/10.3390/antibiotics14020166
12. Arastehfar A, Gabaldón T, Garcia-Rubio R et al (2020) Drug-resistant fungi: an emerging challenge threatening our limited antifungal armamentarium. Antibiotics 9:877. https://doi.org/10.3390/antibiotics9120877
13. Prasad R, Nair R, Banerjee A (2019) Multidrug transporters of Candida species in clinical azole resistance. Fungal Genet Biol 132:103252. https://doi.org/10.1016/j.fgb.2019.103252
14. AlQurashi DM, AlQurashi TF, Alam RI et al (2025) Advanced nanoparticles in combating antibiotic resistance: current innovations and future directions. J Nanotheranostics 6:9. https://doi.org/10.3390/jnt6020009
15. Wu Q, Cen F, Xie Y et al (2025) Nanoparticle-based antifungal therapies innovations mechanisms and future prospects. PeerJ 13:e19199. https://doi.org/10.7717/peerj.19199
16. Devaraji M, Thanikachalam PV, Elumalai K (2024) The potential of copper oxide nanoparticles in nanomedicine: a comprehensive review. Biotechnol Notes 5:80–99. https://doi.org/10.1016/j.biotno.2024.06.001
17. Wang L, Hu C, Shao L (2017) The antimicrobial activity of nanoparticles: present situation and prospects for the future. Int J Nanomedicine 12:1227–1249. https://doi.org/10.2147/IJN.S121956
18. Fayed B (2025) Nanoparticles in the battle against Candida auris biofilms: current advances and future prospects. Drug Deliv Transl Res 15:1496–1512. https://doi.org/10.1007/s13346-024-01749-w
19. Liu Y, Zhang Y, Li H et al (2025) Recent advances in the bench-to-bedside translation of cancer nanomedicines. Acta Pharm Sin B 15:97–122. https://doi.org/10.1016/j.apsb.2024.12.007
20. Đorđević S, Gonzalez MM, Conejos-Sánchez I et al (2022) Current hurdles to the translation of nanomedicines from bench to the clinic. Drug Deliv Transl Res 12:500–525. https://doi.org/10.1007/s13346-021-01024-2
21. Malik MA, Wani AH, Bhat MY et al (2024) Fungal-mediated synthesis of silver nanoparticles: a novel strategy for plant disease management. Front Microbiol 15:1399331. https://doi.org/10.3389/fmicb.2024.1399331
22. Lakhani KG, Hamid R, Motamedi E et al (2025) A review on plant metabolite-mediated nanoparticle synthesis: sustainable applications in horticultural crops. Front Nanotechnol 7:1545413. https://doi.org/10.3389/fnano.2025.1545413
23. Choi JS, Lee JW, Shin UC et al (2019) Inhibitory activity of silver nanoparticles synthesized using Lycopersicon Esculentum against biofilm formation in Candida species. Nanomaterials 9:1512. https://doi.org/10.3390/nano9111512
24. Ratan A, Singh S, Kumar A (2024) Comparative antimicrobial activity of silver nanoparticle synthesized using Lantana camara leaves. Int J Curr Microbiol Appl Sci 13:129–136. https://doi.org/10.20546/ijcmas.2024.1311.015
25. Desai N (2012) Challenges in development of nanoparticle-based therapeutics. AAPS J 14:282–295. https://doi.org/10.1208/s12248-012-9339-4
26. Chaudhari S, Panchave R (2022) Regulatory and toxicology issues surrounding nanomedicines. International Journal of Creative Research Thoughts (IJCRT), 10(1), c255–c260. ISSN: 2320–2882

27. Cheng T-M, Chu H-Y, Huang H-M et al (2022) Toxicologic concerns with current medical nanoparticles. Int J Mol Sci 23:7597. https://doi.org/10.3390/ijms23147597
28. Das D, Ahmed G (2012) Effects of silver nanoparticles on the structural integrity of yeast cell walls. Adv Agric Appl Sci 2(1):1–3. ISSN: 2788–2711
29. Ulaş F, Yüksel E, Dinçer D et al (2025) Recent advances in plant-based green synthesis of nanoparticles: a sustainable approach for combating plant-parasitic nematodes. Sustainability 17:4152. https://doi.org/10.3390/su17094152
30. McCracken E, Monaghan M, Sreenivasan S (2018) Pathophysiology of the metabolic syndrome. Clin Dermatol 36:14–20. https://doi.org/10.1016/j.clindermatol.2017.09.004
31. Ribeiro LG, Roque GSC, Conrado R et al (2023) Antifungal activity of mycogenic silver nanoparticles on clinical yeasts and Phytopathogens. Antibiotics 12:91. https://doi.org/10.3390/antibiotics12010091
32. Seddighi NS, Salari S, Izadi AR (2017) Evaluation of antifungal effect of iron-oxide nanoparticles against different Candida species. IET Nanobiotechnol 11:883–888. https://doi.org/10.1049/iet-nbt.2017.0025
33. Djearamane S, Xiu L-J, Wong L-S et al (2022) Antifungal properties of zinc oxide nanoparticles on Candida albicans. Coatings 12:1864. https://doi.org/10.3390/coatings12121864
34. Slavin YN, Bach H (2022) Mechanisms of antifungal properties of metal nanoparticles. Nanomaterials 12:4470. https://doi.org/10.3390/nano12244470
35. Tijani NA, Afolabi AO, Mutanule F et al (2025) Exploring the anticandida potentials of nanoparticles fabricated from tea leaf (Camellia sinensis) extracts: a systematic review. Discov Mater 5:14. https://doi.org/10.1007/s43939-025-00196-9
36. Hano C, Abbasi BH (2021) Plant-based green synthesis of nanoparticles: production, characterization and applications. Biomolecules 12:31. https://doi.org/10.3390/biom12010031

Chapter 8
Underutilized Plant-Derived Nanoparticles Against Pathogenic Fungi

Esraa E. Ammar, Soumya Ghosh, Ibrahim A. I. Hassan, Mohammed E. Gad, Shrouk Ekramy, Ibrahim W. Ibrahim, and Hesham A. M. Ibrahim

Abstract The use of plant-derived nanoparticles (PDNPs) as a novel solution to fungal infection has been developed based on a recent, alternative, and sustainable approach to extracting substances from underused plant resources. Phytochemicals present in these plants are natural, efficient agents that reduce and stabilize without the aid of toxic or costly chemicals. This is a green strategy that is beneficial. PDNPs show pronounced antifungal effects by various mechanisms, such as interference with fungal cell membranes, oxidative stress induction, and biofilm formation inhibition. The increasing prevalence of drug-resistant fungal infections combined with the toxicity, high cost, and resistance of traditional antifungal medications implements the urgency of the necessity of alternative therapeutic approaches. The use of PDNPs, which is a part of green nanotechnology, offers a promising solution

E. E. Ammar (✉)
Plant Ecology, Botany Department, Faculty of Science, Tanta University, Tanta, Egypt
e-mail: esraa_ammar@science.tanta.edu.eg

S. Ghosh
Natural and Medical Sciences Research Center, University of Nizwa, Nizwa, Oman

University of the Free State, University of the Free State, Bloemfontein, South Africa

I. A. I. Hassan
Department of Chemistry, Qena University, Qena, Egypt

M. E. Gad
Department of Zoology and Entomology, Faculty of Science, Al-Azhar University, Cairo, Egypt

S. Ekramy
Botany and Chemistry Department, Faculty of Science, Al-Azhar University, Cairo Governorate, Egypt

I. W. Ibrahim
Department of Biotechnology, Faculty of Agriculture, Cairo University, Giza, Egypt

H. A. M. Ibrahim
Department of Agricultural Zoology and Nematology, Faculty of Agriculture, Al-Azhar University, Assiut, Egypt

© The Author(s), under exclusive license to Springer Nature Switzerland AG 2026
E. Nepovimova et al. (eds.), *Underutilized Plants Based-Nanoparticles: Herbal Remedy to Treat Human Diseases*, https://doi.org/10.1007/978-3-032-20384-7_8

because it is more biocompatible, less toxic, and has fewer chances of developing resistance. Moreover, the importance of preserving the underutilized plant species through valorization not only helps to discover new bioactive substances but also encourages the preservation of biodiversity and sustainable use of resources. The chapter brings out the modern developments, contemporary issues, and future outlooks of PDNPs as second-generation antifungal agents to be used both in agriculture and healthcare.

Keywords Pathogenic yeast · Nanoparticles · Underutilized plant-derived · Economical sustainable and environmentally friendly approaches

8.1 Introduction

Fungi can be found in an unbelievable variety of environments, such as the very high atmosphere in the stratosphere, and the extreme arid climate of deserts, as well as deep in the ocean bed, in the ice fields of the Antarctic glaciers, and in the microbial community of the human gastrointestinal tract. Though comprising less than 1% of the gut microbiota, there is increasing evidence of their vital role in the regulation of immune processes, and the establishment of interactions with the host who are not only symbiotic but are often also mutualistic. According to epidemiological evidence, there is a significant health burden of fungal diseases across the world. Over 2.1 million patients are reported to have invasive aspergillosis alone annually, including many who subsequently develop chronic obstructive pulmonary disease, malignancies, or hematological disorders, and with this, there will be about 1.8 million deaths, which is equivalent to 85.2% death rate Persistent aspergillosis of the lungs is responsible for an estimated 1,837,272 cases each year, leading to approximately 340,000 deaths, representing a mortality rate of about 18.5%. Annually, Candida-related bloodstream infections affect nearly 1.56 million individuals, resulting in around 995,000 fatalities (about 63.6% mortality). Similarly, Pneumocystis pneumonia accounts for roughly 505,000 cases and 214,000 deaths (mortality rates of 42.4% and 75.8%, respectively). Meanwhile, cryptococcal meningitis contributes to approximately 194,000 infections and 147,000 deaths worldwide. Other severe and life-threatening fungal diseases collectively affect between 300,000 and 400,000 individuals annually, with an estimated 161,000 deaths (around 53.7%). Additionally, it is estimated that 11.5 million individuals worldwide suffer from fungal asthma, which causes 46,000 asthma-related deaths annually. Overall, an estimated 6.5 million invasive fungal infections occur worldwide each year, resulting in 3.8 million deaths, of which almost 2.5 million (68%) are directly linked to fungal pathogens [1].

The problem of antifungal resistance (AFR) came as an urgent issue on the global health agenda with dire effects, especially in the frail-immune population. A One Health approach is necessary to address AFR. That includes Stewardship of

Antifungals (AFS) as one of its core elements. Worldwide, fungal infections cause about 150 million cases annually, and this causes about 1.7 million deaths annually. The duration of AFR is also a major problem as the infections by anti-fungal resistant strains are in fact known to be abnormally difficult to treat [2]. The most recent example of recognizing fungal pathogens as global health priorities has been brought up by the World Health Organization (WHO), where antifungal resistance (AFR) is one of the key emerging public health threats on a global scale. The environmental conditions are closely connected with AFR and fungal epidemiology because most clinically relevant species of fungi, along with their resistant strains, begin in natural sources of infection, particularly soil [3]. The fungus can transfer to new sites in the environmental changes, and this can expose the human populations to new strains of fungi, which are more virulent or resistant to drugs. The combined forces contribute to the likelihood of infections of a more aggressive virulence profile, an augmented resistance to antifungal treatment, and completely novel dangers of fungi. The growing menace highlights the need for multidisciplinary actions in which the establishment of policies combined with educating the general population and healthcare experts, novel antifungal therapy development, enhanced diagnostics during the early stages, and the minimization of healthcare disparities are all implemented. Such efforts have a significant impact on lowering the incidence of fungal diseases worldwide.

Plant-mediated synthesis of nanoparticles has become a green, sustainable, and ecologically friendly method in recent decades. During the creation of nanoparticles, this green synthesis uses plant-derived extracts as stabilizing or capping agents in addition to reducing agents, presenting a host of benefits including enhanced safety, diminished environmental footprint, as well as increased resource efficiency. The ability of plants to catalyze the synthesis of nanoparticles is due to high-density, which contains phytochemicals such as terpenoids, flavonoids, polyphenols, and other bioactive substances. Minimization of toxic reagents decreases environmental pollution and health hazards related to classical synthesis routes. In addition, native bioactivity of plant extracts may impart novel functional features to nanoparticles, enhancing applicability towards medicine, remediation of pollutants, and industrial practices. However, variation in plants' species, cultivation conditions, and extract preparation may result in variance in nanoparticle diameter, morphology, and functionalities. To address such setbacks, standardized characterization methodologies are needed [4].

8.2 Underutilized Plants: Untapped Reservoirs of Bioactive Phytochemicals

Underutilized plants are those species that are least available, out of use, or are locally confined. Such species hold great promise but are largely overlooked by plant breeders, researchers, as well as policymakers. The description itself of being underutilized carries neither any geographical, social, nor economic dimension related to plants. Various terms are used interchangeably to describe them, including orphan crops, neglected crops, indigenous crops, traditional crops, minor crops, and sometimes even poor people's food. Underutilized plants are typically classified according to their end-use, including categories such as beverages, cereals, nuts, vegetables, fruits, and medicinal plants. Several indigenous leafy vegetables have been reported to hold significant nutritional potential that could contribute positively to the diets of rural populations. Dark green leafy vegetables, while rich in nutrients, are also known to contain anti-nutritional factors such as oxalates, phytates, nitrates, tannins, and phytohemagglutinin agents [5].

The most reported medical applications were for the treatment of skin conditions, respiratory issues, and gastrointestinal disorders. In most cases, topical applications came after oral treatment. Although every part of the plant was used, the leaves were the most frequently used. Zlatibor district's folk medicine, as summarized in South-Western Serbia, is primarily meant to be used as a primary healthcare method for the treatment of minor ailments. The findings show a minor decline in this field's ethnobotanical and medicinal expertise in comparison to adjacent areas [6]. In South-Western Serbia, there are 69 species of plants from 36 families used as medicinal medicines. The three largest families of plants were *Asteraceae* (12%), *Lamiaceae* (13%), and *Rosaceae* (16%). In this study, three or more informants mentioned 37 different plant species. Species cited by just one or two informants should not be discounted, even if some studies utilize this criterion to assess the legitimacy of medical plant usage, as their traditional knowledge may have declined over time, indicating the need for further investigation. Based on the use-value (UV) calculation, the species with the highest usage values were *Mentha piperita* (0.905), *Matricaria chamomilla* (0.686), *Hypericum perforatum* (0.541), and *Achillea millefolium* (0.268) [7, 8]. Table 8.1 lists the medicinal uses of the plants that were examined. Treatment of respiratory and gastrointestinal disorders accounted for 33.06% of use-reports, making them the most reported medical purposes. (29.56%), dermatological conditions (14.83%), urinary tract disorders (6.08%), circulatory system issues (5.15%), neurological and psychiatric disorders (5.15%), nutritional issues and endocrine gland and metabolism disorders (4.33%), issues with the skeleton, muscles, and connective tissues (1.03%), gynecological complications (0.51%), and sensory system issues (0.30%). The importance of using medicinal plants in the gastrointestinal tract.

Compared to widely used therapeutic plants, underutilized plants offer prospective sources of bioactive secondary metabolites that have not yet been thoroughly investigated [23]. These substances, which have anti-inflammatory and anti-cancer plus anti-microbial, antidiabetic, additional to antioxidant activity, include flavonoids, terpenoids, alkaloids, polyphenols, tannins, and saponins [24]. More than 50,000 plant-derived chemicals have been chemically described to date, according to a recent analysis, and fresh findings are made every year [24, 25]. Plants represent a great repository of varied bioactive compounds with great therapeutic promise. *Calotropis procera* possesses anticancer and free radical-reducing potential due to its concentrated content of flavonoids, steroids, and volatile oils [26]. Similarly,

Table 8.1 Underutilized plants and their usage in medicine for diseases and their formulation

Scientific names	Medicinal uses	Preparation method and extraction	Reference
Acacia modesta Wall.	Backache	Halwa, a remedy for backaches, is made by combining wheat flour, desi ghee, and the gum and powder of fresh Acacia modesta leaves	[9]
Acacia nilotica (L.) Delile	Narcotic Aphrodisiac Earache	Fresh root and leaf extraction is used to make alcohol (Sharab) 4 g of gum are mixed with water to form a paste 10 mL of mustard oil are used to heat and filter approximately 30 flowers	[9]
Ficus carica L.	Antihemorrhoidal Anticatarrhal Warts	Decoction Decoction (with wine)	[10]
Phillyrea angustifolia L.	Injuries Ulcers in the mouth	Infusion Decoction	[11, 12]
Clematis campaniflora Brot.	Pneumonia	Decoction	[13]
Acorus calamus L.	Digestive problems Chronic dysentery	Dry the roots to be powdered and mix it with sugar and take by oral	[14]
Achyranthes aspera L.	High fever Chest diseases	Dry the lower stem and mix with powdered leaves of *Mentha longifolia* and taken by oral	[15, 16]
Amaranthus spinosus L.	Biliousness	Leaves and small shoots is boiled in water and their decoction is used	[14, 17]
Bergenia ciliata (Haw.) Sternb.	Urinary problems Skin infections Demulcent	Decoction of the rhizome and taken 3 times a day	[18, 19]
Asystasia schimperi T. Anders	Cough Skin diseases	Infusion	[20]
Periploca linearifolia Dill. & rich	Venereal diseases- Warts- Rituals-Pneumonia-Cancer-Antidiarrhea-fertility	Internal decoction and exudates	[21]
Solanecio mannii (Hoof.f.) C. Jeffrey	Typhoid, cough, cancer, pneumonia, and epilepsy	Decoction	[22]

mangroves like Avicennia and Rhizophora are great reservoirs of polyphenols with established antiviral and anticancer activities [27]. Moringa oleifera is yet a remarkable plant with a broad range of bioactive compounds with antibacterial, free radical scavenging, and nutritive value. Characteristic of such phytochemicals is that they possess a capability of being multi-target therapeutic agents in that they adjust major signaling cascades such as MAPK, NF-κB, as well as PI3K/Akt [28, 29]. Such an observation underscores a great promise of underutilized plants as a great underexploited resource for drug development, presenting a chance of discovering new bioactive molecules with potency to treat infectious diseases, cancer, as well as chronic ailments [28, 29].

8.3 Green Synthesis of Nanoparticles Using Underutilized Plants

The foundation of green chemistry was established by Warner and Anastas who introduced the concept in their seminal 1998 book, "*Green Chemistry: Theory and Practice*." In this work, they proposed 12 guiding principles designed to help scientists, industry professionals, and policymakers minimize or cut back on operations that are detrimental to the environment and stop using dangerous materials. The values and their visibility, practicability, and applicability have been core to the development of green chemistry as a scientific approach, which is concerned with the fabrication of more sustainable and less hazardous chemical processes and products which in their turn protect human health and the environment [30]. We must avoid the use of other chemicals that could be hazardous in case they could be of no benefit. Design to Save Energy: Reduce uses of energy in the chemical synthesis. Renewable Feedstocks: Use renewable feedstocks and make sure that their use does not cause a depletion of resources. Minimize Derivatives: Do not use derivatives, including blocking agents or protecting/deprotecting groups because they cause unnecessary waste. Catalysis: Compared to stoichiometric agents, catalysis agents are better. Design to Degradation: At the conclusion of the synthesis process, chemicals should be made to produce non-toxic derivatives. Pollution Prevention Analysis: Use a real-time analysis of the production of toxic chemicals during the synthesis process to avoid pollution. Inherently Safer Chemistry to Prevent Accidents: Choose reagents and processes that reduce the risk of hazardous incidents during chemical production. The reduction of metal complexes in diluted solutions is the fundamental idea behind environmentally acceptable methods for creating metal oxide nanoparticles. The process's objective is to create metal colloidal dispersions. Plant-based or other organic sources are rich in specific molecules capable of converting metal precursors into nanoparticles, which has driven significant interest in plant extracts. Additionally, the green synthesis approach not only facilitates nanoparticle formation but can also coat the nanoparticles in situ, using natural materials from plants, algae, fungi, or microorganisms as stabilizing & reducing agents. By taking part in the creation process, they lessen toxicity, enhance the antimicrobial activity of the nanomaterial, and inhibit the agglomeration of nanoparticles, maybe producing a synergistic effect [31] (Fig. 8.1).

As Chakraborty et al. (2022) and his colleagues [32] said 10 mL of 1.0 mM silver nitrate solution was mixed with four different amounts (1.0, 2.0, 3.0, 4.0 g) of the root extract of *K. angustifolia* plants grown both in vitro and in vivo. The solutions were shaken at 80 rpm and incubated for 10 h at 45 °C. The nanoparticles were then

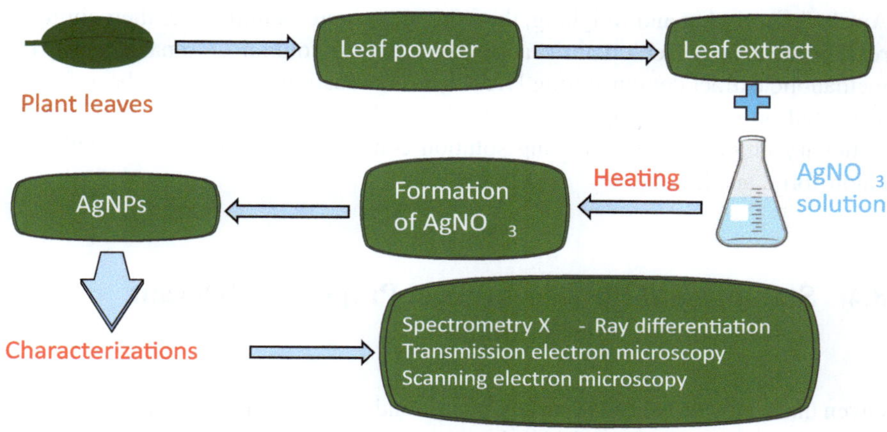

Fig. 8.1 A picture of how silver nanoparticles derived from plants are made. The technique of creating green synthesis nanoparticles with silver components is depicted in this image

incubated for a further 10 h at 28 °C to achieve the proper size distribution. Silver bio-nanoparticles were created when the solutions gradually changed from translucent to bright yellow to light brown to dark brown. The nanoparticles were then preserved in dry, cool conditions. Therefore. Utilizing leaf extract from the *C. fistula* plant as fuel, ZnO nanoparticles were created by an environmentally benign green combustion method. The hydrated zinc nitrate (Zn $(NO_3)_3$ $6H_2O$) without purification in 10 mL of distilled water, 0.2 g of *C. fistula* leaf extract was used to dissolve a stoichiometric quantity of (Zn $(NO_3)_3$ $6H_2O$).

The mixture was then left to burn in a preheated muffle to 400 + 10. After 5 min, a fine, milky-white substance was produced. The various plant extract quantities (0.3, 0.4, 0.5, and 0.6 g) were used in the creation of the nanoparticles. Until it was needed again, the product was kept in an airtight container [33]. Terpenoids' carbonyl group in *Curcuma mangga* extract was discovered to disulfide Au 3 to Au 0 and create nuclei, which further combine to form colloidal Au NPs. Zerumin A, Coronadiene, and Longpene A are terpenoids whose carboxylic groups adsorb on the surface of NP [34]. Ag colloids are formed because of the stabilizing action that causes them to aggregate further and form oligomeric clusters. Ag NPs are stabilized and reduced by oxidized polyphenolic terpenoids through the C=O bond. A. Terpenoid fractions. Paniculate leaf extract serves as a mediator in the Zn^{2+} reduction to ZnO, yielding colloidal ZnO NPs. SPR vibrations may be the cause of color change observed in the extract upon Zn $(NO_3)_2$ treatment. 1 g of dried *Psidium guajava* (Pg) leaf biomass was dissolved in 100 mL of a 1×10^{-3} M aqueous $HAuCl_4$ solution, where $AuCl_4^-$ acted as the bioreductant, in order to create gold nanoparticles utilizing guavanoic acid [35]. The reaction combination UV-vis spectra with a resolution of 1 nm were recorded on solution-cast films of gold nanoparticles on a quartz substrate. 800 g of Pg leaf biomass was extracted using methanol, chloroform, ethyl acetate, acetone, petroleum ether, and hexane separately in a Soxhlet system until exhaustion to characterize the molecule. That reduced and capped

$AuCl_4$. After drying and weighing, the extractions were examined for their ability to react with aqueous chloroauric acid. By utilizing column chromatography, the methanolic extract demonstrating $HAuCl_4$'s bioreduction property has been further separated into seven components (60–120 mesh). A glass column is used for the stationary phase, and a dissolving solution consisting of a gradient of methanol, chloroform, and hexane.

8.4 Stability and Physicochemical Properties Relevant to Antifungal Applications

Given that characteristics like surface charge and monodisperse particle size directly affect efficacy (zeta potential) that is applicable to the situation, hydrophobic balance, and colloidal stability (measured by hydrodynamic size and polydispersity index), stability and physicochemical characteristics of nanoparticles are important in determining efficacy by controlling interactions between nanoparticles and fungal cells. These nanoparticles can be considered as suitable for a diverse range of applications such as crop protection in agriculture, topical and systemic therapy of human infections due to such multiple antifungal actions, including membrane disruption, production of reactive oxygen species (ROS), DNA damage, and enzyme inhibition. Some issues that need to be overcome include host cell toxicity, aggregation, lack of selective specificity, danger of resistance development, and scalability. Possible solutions are strategic surface modification, the creation of multifunctional nanoparticles for targeted delivery and controlled release, combination therapy with conventional antifungals, and green synthesis methods. Future directions remain dedicated to personalized medicine advancement, nanoparticle-based diagnostics creation, enhanced targeted delivery systems, combination nanoparticles and immunotherapy, and advanced novel material exploration to combat antifungal drug resistance and improve treatment outcomes.

8.5 Mechanisms of Antifungal Action

Currently, several antifungal drug classes are available for treatment. Polyenes, such as amphotericin B (AmB), act by disrupting fungal cell membranes, resulting in cell death and structural damage. Although some species, such as *Cryptococcus*, are resistant, 5-fluorocytosine (5-FC) suppresses nucleic acid synthesis, particularly when administered in combination. Although certain *Candida* species show resistance, echinocandins, such as micafungin, target the formation of fungal cell walls. Although resistance in *Candida* species is also frequent, azoles, like fluconazole, interfere with fungal ergosterol production, impairing membrane integrity and metabolism [1]. A prodrug called 5-fluorocytosine (5-FC) enters cells via the

Table 8.2 An overview table contrasting antifungal medications, their modes of action, and their clinical uses

Classes of ligand	Ligand	Mechanism	Clinical approaches	References
Azoles	Fluconazole, itraconazole, and voriconazole	Target lanosterol 14α-demethylase to prevent ergosterol synthesis and cause membrane uncertainty	Candidiasis, aspergillosis, cryptococcosis	[39]
Echinocandins	Caspofungin, micafungin, additional to anidulafungin	Impair the integrity of fungal cell walls by inhibiting β-glucan synthase	Invasive candidiasis and aspergillosis	[40]
Polyenes	Amphotericin B and nystatin	Bond to ergosterol, which increases the permeability of membranes and kills cells	Systemic fungal infections and oral and vaginal candidiasis	[41]
Allylamines	Terbinafine and naftifine	Stop the production of ergosterol by inhibiting squalene epoxidase	Dermatophytoses (e.g., ringworm and athlete's foot)	[42]
Pyrimidine analogs	Flucytosine	Transformed into 5-fluorouracil, which interferes with the formation of fungal proteins and RNA	Cryptococcal meningitis and candidiasis (used in combination therapy)	[43]
Griseofulvin	Griseofulvin	Inhibits mitosis by interfering with microtubule activity	Dermatophytoses (e.g., tinea infections)	[44]

cytosine permease gene (FCY2), where it prevents the production of nucleic acids, hence impeding the growth of fungi. Relevant diagrams show the precise mechanisms of action of certain antifungal drugs [36, 37]. According to a previous study [38], conventional antifungal medications fall into roughly five categories: azoles, polyenes, echinocandins, allylamines, and pyrimidine analogs, which are employed for both topical and systemic antifungal therapies (Table 8.2). The growing health risks posed by fungal infections necessitate the development of effective and reliable antifungal therapies. Nevertheless, the existing clinical repertoire has been reduced to several larger classes: azoles, echinocandins, polyenes, and the antimetabolite flucytosine. A decrease in efficacy, some drug resistance, and side effects remain obstacles to their utility. Furthermore, the increasing of the resistant pathogenic fungi, such as Candida auris, is a critical health issue of the global level, which is why the development of new antifungal agents with better efficacy and less toxicity is urgent [45]. In addition, triazole antifungals like fluconazole work by inhibiting the cytochrome P450 14a-demethylase that converts lanosterol to ergosterol, which is a crucial enzyme called constituent of the cell membrane for fungi. This interference affects the integrity of membranes and heightens cellular permeability, which eventually affects fungal growth in an adverse way. The other antifungal agents act in other ways, including preventing the creation of lanosterol or protein synthesis or disrupting the production of ergosterol. Also, amphotericin B (AmB) acts by creating holes in the fungal cell wall, which results in the loss of ions and small molecules that cause cell lysis and eventual death of fungi [1]. Antifungal resistance is a multidimensional and multifactorial issue that occurs in the topography of multiple biological and environmental factors interaction. The proximate causal factors are: (1) increased or enhanced expression or up-regulation of the drug target; (2) the ad hoc resistance of novel antifungal drugs; (3) adherence to prescribed regimens; (4) inappropriate or excessive use of antifungal agents; (5) serendipitous genetic changes in fungi strains; (6) the poor management and disposal of pharmaceutical wastes; (7) altered or altered metabolic pathways, which decrease the efficacy of antifungal factors [46]. The current issue of antifungal resistance is growing, hence there is a need to proceed with new therapeutic agents in all antifungal drug classes through continuous research and development. The ideal antifungal compound must contain a number of desirable properties: (1) low or well-tolerated toxicity with minimal side effects, which would enable it to have a broad therapeutic index; (2) good pharmacokinetics that would enable the compound to be used via a variety of routes of administration; (3) selectivity in actions against fungi cells, but not host tissue; (4) the compound should preferably be fungicidal but not fungistatic; (5) it should have a wide range of action against a various fungi pathogens, the use of natural bioactive molecules, semisynthetic and synthetic derivatives, nanomaterials, peptides, as well as agents with a novel biochemical pathway is among the current and emerging directions in antifungal discovery [47]. Thus, it is essential to get a more extensive insight into the mechanism of antifungal activity of natural products. Plant-derived antifungal compounds usually affect a variety of processes, including cell wall or membrane integrity, oxidative stress generation through reactive oxygen species (ROS), biofilm formation, metabolic activity, and gene expression.

8.6 Inhibition of Biofilm Formation

In case of invasive infections, fungal cells adhere to the skin surface or the mucosa and form polysaccharide-based biofilms, which are very difficult to clear. This complexity of organization is a reason for drug resistance and enhanced immune evasion. Though numerous agents inhibit biofilm formation, very few antifungal medications are available, and they are mostly associated with toxicity [48, 49]. Therefore, natural products, the antibiofilm properties of plants, especially those employed in traditional medicine, have drawn attention. For example, when tested against *Candida* planktonic cells, the phytochemicals carvacrol, cinnamaldehyde, and thymol dramatically reduced mature biofilm biomass and metabolic activity, whereas cinnamaldehyde hindered biofilm adhesion [50]. Similarly, 6-shogaol of *Zingiber officinale* (ginger) exhibited antibiofilm activity by both inhibiting biofilm formation and eradicating preformed *C. auris* biofilms and decreasing efflux pump-associated CDR1 gene [51]. Moreover, warburganal and polygodial isolated from *Warburgia ugandensis* exhibited remarkable anti-biofilm activity of *C. albicans*; upon addition of azoles, the compounds inhibited biofilm formation approximately ten times compared to single treatments [52].

Azole antifungals fluconazole, voriconazole, and itraconazole are generally ineffective against Candida biofilms, although amphotericin B remains one of the few agents to significantly suppress their formation [37]. Biofilms, groups of sessile fungal cells that develop on the surface, exhibit enhanced resistance by mechanisms such as extrusion of growth rate variations, resistance gene expression, and extracellular polymeric substances (EPS). These are factors related to drug tolerance and pathogenicity, which constitute an important clinical issue, particularly with *C. albicans*. *C. albicans* species possess more complex biofilms with a robust extracellular matrix (ECM) structure compared to *S. cerevisiae*, although both yeast and filamentous cells can form biofilms. Ag-doped hydroxyapatite nanoparticles (Ag-HA NPs) inhibited *C. albicans* biofilm biomass, suggesting disruption of ECM synthesis. The inhibition did not translate to lower colony-forming unit (CFU) counts, suggesting possible recovery of biofilms. Surprisingly, sub-inhibitory NP levels at times stimulated growth of biofilms, possibly due to upregulation of virulent genes. Exposure to AgNPs reduced biofilm biomass as well as CFU counts in *C. albicans* 24 and 48 h, with higher efficacy observed when it was tested against attached cells as opposed to mature biofilms [53]. When invading the host, fungal pathogens are often engulfed by immune cells such as macrophages. Pathogen–phagocyte interaction studies show that intracellular conditions induce persister cell formation. Comparative metabolic profiling of planktonic and persister cells challenged with oxidative or pH stress indicates that stress responses play a part in persistence. Although biofilm formation is not required for the formation of persisters, biofilm-associated persisters are more resistant to oxidative damage and resist amphotericin B [54]. Biofilm inhibition may therefore enable the killing of *C. tropicalis* persister cells. Antifungal agents tend to enhance levels of ROS, and hence persisters can boost antioxidant defense mechanisms by inducing such enzymes as superoxide dismutases (SODs) and alkyl hydroperoxide reductase 1 [36]. Berberine (BBR), a medicinal isoquinoline alkaloid found in *Coptidis rhizoma*, *Berberis* spp., and *Hydrastis canadensis*, possesses broad-spectrum antifungal and antibiofilm activity, effectively suppressing biofilm growth-related resistance [53, 55]. Recent developments in quorum-sensing inhibitors (QSIs) can prevent the formation of biofilms in unicellular eukaryotes such as *Candida albicans* and related organisms are presented. The table outlines each QSI's mechanism of action, targeted organism, and reported reduction in biofilm biomass or viability. The variety of research in this area is demonstrated by the representation of both synthetic inhibitors and natural products (such as garlic extract), and their potential for improved biofilm disruption in the clinical setting is highlighted by synergistic approaches with conventional antifungals using QSIs [56].

Enzymes called β-glucanases hydrolyze the structural polysaccharides of the biofilm matrix, weakening them and increasing the accessibility of microbial cells to antifungal substances. Mannosidase Inhibitors: Mannan, a noteworthy polysaccharide in the (ECM) extra-cellular dialect, plays a part in the maintenance of biofilms in Candida species. Biofilm biomass is reduced and antifungal activity is increased when mannosidase inhibitors block the synthesis of mannosidase and the integration of these polysaccharides into the biofilm matrix. Enzymes that break down proteins: The extracellular matrix proteins of the biofilm, hydrophobins, and adhesins are essential for the cohesion and adherence of the biofilm. It has been discovered that these extracellular matrix proteins are potent biofilm disruptors by breaking down proteolytic enzymes. Consequently, the effects on the stability and persistence of biofilms: The extracellular matrix (ECM), which protects *C. albicans* from external stress, mediates the biofilm's stability and persistence. Biofilms are susceptible to antimicrobial therapy when they are destabilized by agents that target the extracellular matrix. Defective Biofilm Cohesion: ECM, which is composed of both protein-based and polysaccharide-based ECM, is also in charge of biofilm cohesion. Any disruption to this matrix could lead to a loss of structural integrity, leaving biofilms vulnerable to therapies [56].

8.7 Interference with Fungal Metabolic Pathways and Gene Expression

One of the main targets of antifungal drugs is ergosterol production, which is essential to fungal cell survival. Antifungal resistance could result from this pathway's mutation. For instance, gain-of-function mutations in the transcriptional regulator Upc2 or an increased number of copies of chromosome 5 containing ERG11 are responsible for the overexpression of the gene ERG11, which reduces the susceptibility to azole. Clinical isolates of azole-resistant *C. albicans* frequently have ERG11 mutations. Certain natural substances, including apigenin, rutin, and palmitic acid, have been shown to downregulate ERG11. Other studies have concentrated on the production of ergosterol using cold atmospheric plasma, tormentic acid, benzodioxane analogs, honokiol, and magnolol [57]. Ergosterol is not the only metabolic route that can be targeted by antifungals. Among other vital cellular functions, these include the metabolism of N-acetylglucosamine, trehalose, and lipids. Critical fungal enzymes are also appealing targets for medications because of their vital roles in virulence and proliferation. For example, the antimicrobial peptide AMP-17 disrupts the main metabolic cascades of *C. albicans*, including oxidative phosphorylation, RNA degradation, fatty acid metabolism, and propanoate metabolism. *Aspergillus niger* spores' cell membrane integrity and metabolic activity have been found to be impacted by the apolipoprotein B (ApoB) peptides. Animals and higher plants do not contain the class II fructose-1,6-bisphosphate aldolase (FBA-II) enzyme, which is another promising target and potentially selective. Phenylhydrazone derivatives have also demonstrated strong inhibitions of C. albicans FBA-II, with an IC50 of less than 10 μM. According to tests using molecular docking, molecular dynamics, and enzyme inhibition assays, thiazolidinones inhibit *Candida glabrata* carbonic anhydrase (KI = 0.1 10 μM). Also, compounds like vanillin and perillyl alcohol have been reported to inhibit the glyoxylate cycle in *C. albicans*. Other natural or semi-synthetic molecules also target specific fungal enzymes. For example, α-bisabolol inhibits Δ24-sterol methyltransferase, reducing ergosterol synthesis by 26–74% in a dose-dependent manner and suppressing *ERG6* expression by 76% at 9 mM. The homoserine dehydrogenase inhibitor (S)-2-amino-4-oxo-5--hydroxypentanoic acid, isolated from *Streptomyces*, acts as a suicide inhibitor of Hom6p and shows activity against *Aspergillus* species, but not against *C. albicans*, *Candida tropicalis*, or *C. glabrata* [57].

Nanoparticle exposure can dramatically change the expression of fungal genes and protein levels (NPs). Similar effects are anticipated when fungal cells meet substances that have antimycotic activity since, as was previously mentioned, changes in gene regulation can contribute to antifungal resistance. Numerous investigations have investigated how NPs affect *Saccharomyces cerevisiae's* genome. According to Horstmann et al., exposure to AgNPs impacted 1845 genes, of which 1077 were upregulated and 768 were downregulated. Nitrogen compound metabolism was linked to 651 of the elevated genes; Other categories included translation, translational start, ribosome biogenesis, and rRNA processing. In particular, rRNA processing-related DBP2 and FAF1 were elevated 7.9 and 7.2 times, respectively [54]. Efflux pumps, particularly members that aggressively export medications from fungal cells, the ABC transporter and major facilitator superfamily (MFS) contribute significantly to antifungal resistance. Overexpression of these pumps can offer a transient survival benefit in heteroresistant subpopulations. Aneuploidy was significantly more common in heteroresistant and recurring isolates of clinical *Cryptococcus neoformans* strains, primarily chr1 disomy, along with elevated efflux pump activity, according to genomic analysis. Fluconazole-induced heteroresistance was largely caused by overexpression of the PDR11 efflux pump. Additionally, fluconazole treatment elevated mRNA levels of ABC transporter genes (CDR family), further promoting heteroresistance [36]. Similarly, treatment with scopoletin significantly affected gene transcription in *Botrytis cinerea*. 3495 differentially expressed genes (DEGs) were found, comprising 2210 downregulated and 1285 upregulated genes. Additionally, scopoletin damaged the fungus's cell walls and membranes, interfered with ROS homeostasis, and adversely controlled the HOG pathway [58].

8.8 Conclusion

In brief, plant-derived nanoparticles (PDNPs), which are made from underused plant species, offer a revolutionary and sustainable approach to the fight against harmful yeasts and fungi. Their affordable and environmentally friendly synthesis, which makes use of these plants' abundant phytochemical reservoirs, avoids the risks associated with conventional chemical procedures while providing a strong, diverse antifungal action. Compared to traditional antifungals, these frequently have better biocompatibility and a lower chance of resistance. They also disrupt membranes, produce ROS, and prevent biofilms. The investigation of PDNPs promotes biodiversity conservation and the immediate therapeutic potential of natural resources in agricultural and medical contexts, as well as their sustainable use. Therefore, using underused plants to synthesize nanoparticles offers a potent, two-pronged approach: creating next-generation antifungal treatments while advancing ecological sustainability and financial feasibility.

Acknowledgments The authors acknowledge Tanta University, Qena University, Al-Azhar University, and Cairo University, Egypt for their support with this work and would also like to acknowledge the support of the University of Nizwa, Oman; Jiangshu University, China; and the University of the Free State, South Africa.

Competing Interest The authors declare that they have no known competing financial interests or personal relationships that could have appeared to influence the work reported in this paper.

Ethical Approval Not applicable.

Author's Contribution **Esraa E. Ammar:** Conceived the concept of this article, data curation, formal analysis, funding acquisition, investigation, methodology, project administration, resources, software, validation, writing of original draft, review and editing.

Soumya Ghosh and Hesham A. M. Ibrahim: Data curation, formal analysis, funding acquisition, investigation, methodology, project administration, resources, software, validation, writing of original draft, review and editing. **Xiaobo Zou:** Project administration, review and editing.

Mohammed E. Gad, Shrouk Ekramy, Ibrahim W. Ibrahim, Ibrahim A. I. Hassan and Nouran A. EL-Shershaby: Formal analysis, investigation, methodology, project administration, resources, validation, writing of original draft.

Funding This manuscript did not receive any funding.

References

1. Wu Q, Cen F, Xie Y et al (2025) Nanoparticle-based antifungal therapies innovations mechanisms and future prospects. Peer J 13:e19199. https://doi.org/10.7717/peerj.19199
2. Mudenda S (2024) Global burden of fungal infections and antifungal resistance from 1961 to 2024: findings and future implications. Pharmacol Pharm 15(4):81–112. https://doi.org/10.4236/pp.2024.154007
3. Yan ZZ, Hu HW, Xiong C et al (2025) Environmental microbiome, human fungal pathogens, and antimicrobial resistance. Trends Microbiol 33(1):112–129. https://doi.org/10.1016/j.tim.2024.08.003
4. Abegunde SM, Afolayan BO, Ilesanmi TM (2024) Ensuring sustainable plant-assisted nanoparticles synthesis through process standardization and reproducibility: challenges and future directions – a review. Sust Chem One World 3:100014. https://doi.org/10.1016/j.scowo.2024.100014

5. Okigbo RN, Anyaegbu CF (2021) Underutilized plants of Africa. J Biol Nat 13(2):34–49. https://doi.org/10.1038/d44148-025-00183-5
6. Šavikin K, Zdunić G, Menković N et al (2013) Ethnobotanical study on traditional use of medicinal plants in South-Western Serbia, Zlatibor district. J Ethnopharmacol 146(3):803–810. https://doi.org/10.1016/j.jep.2013.02.006
7. Johns T, Kokwaro JO, Kimanani EK (1990) Herbal remedies of the Luo of Siaya District, Kenya: establishing quantitative criteria for consensus. Econ Bot 44(3):369–381. https://doi.org/10.1007/BF03183922
8. Le Grand A, Wondergem PA (1987) Les phytotherapies anti-infectieuses de la foretsavane, Senegal (afrique occidentale) I. Un inventaire. J Ethnopharmacol 21(2):109–125. https://doi.org/10.1016/0378-8741(87)90122-X
9. Khan I, AbdElsalam NM, Fouad H et al (2014) Application of ethnobotanical indices on the use of traditional medicines against common diseases. Evid Based Complement Alternat Med 2014(1):635371. https://doi.org/10.1155/2014/635371
10. Srivastav A, Srivastav Y, Hameed A et al (2024) Prevention and cure of dermatology disorders using herbal medications: summary. Int J Indig Herbs Drugs:1–14. https://doi.org/10.46956/ijihd.v9i1.568
11. Agnihotri A, Kaur A, Arora R (2020) Oral ulceration and Indian herbs: a scoping review. Dent J Adv Stud 8(03):071–079. https://doi.org/10.1055/s-0040-1716316
12. Motti R, de Falco B (2021) Traditional herbal remedies used for managing anxiety and insomnia in Italy: an ethnopharmacological overview. Horticulturae 7(12):523. https://doi.org/10.3390/horticulturae7120523
13. González JA, García-Barriuso M, Amich F (2010) Ethnobotanical study of medicinal plants traditionally used in the Arribes del Duero, Western Spain. J Ethnopharmacol 131(2):343–355. https://doi.org/10.1016/j.jep.2010.07.022
14. Rajput SB, Tonge MB, Karuppayil SM (2014) An overview on traditional uses and pharmacological profile of Acorus calamus Linn. (Sweet flag) and other Acorus species. Phytomedicine 21(3):268–276. https://doi.org/10.1016/j.phymed.2013.09.020
15. Akbar S (2020) Handbook of 200 medicinal plants: a comprehensive review of their traditional medical uses and scientific justifications. Springer International Publishing. https://doi.org/10.1007/978-3-030-16807-0
16. Srivastav S, Singh P, Mishra G et al (2011) Achyranthes aspera-an important medicinal plant: a review. J Nat Prod Plant Resour 1(1):1–14. www.scholarsresearchlibrary.com
17. Kapadia MD, Kharat AR (2012) Antidiarrhoeal activity of leaves of Acorus calamus. Int J Pharm Sci Res 3(10):3847–3850. https://doi.org/10.13040/IJPSR.0975-8232.3(10).3847-50
18. Hashim S, Jan ASAD, Marwat KB et al (2014) Phytochemistry and medicinal properties of Ammi visnaga (Apiacae). Pak J Bot 46(3):861–867
19. Kour H (2019) Studies on wound healing potential of Bergenia ciliata rhizome. Doctoral dissertation, Kashmir University
20. Ruth L, Manani Solomon D (2010) Ethnobotanical survey and propagation of some endangered medicinal plants from south Nandi district of Kenya. J Anim Plant Sci 8(3):1016–1043. http://www.biosciences.elewa.org/JAPS
21. Belay WY, Gurmu AE, Wubneh ZB (2022) Antimalarial activity of solvent fractions of the stem bark of Periploca linearifolia Quart.-Dill. & A. Rich. (Asclepiadaceae) against plasmodium berghei in mice. J Exp Pharmacol 14:7–15. https://doi.org/10.2147/jep.s328954
22. Jeruto P, Lukhoba C, Ouma G et al (2008) An ethnobotanical study of medicinal plants used by the Nandi people in Kenya. J Ethnopharmacol 116(2):370–376. https://doi.org/10.1016/j.jep.2007.11.041
23. Macha MA, Zargar SM, Bhat AA et al (2025) Neglected and underutilized crops: unveiling potential for cancer drug discovery, 1st edn. CRC Press. https://doi.org/10.1201/9781003477792
24. Hassan S, Ganai BA (2025) Neglected and underutilized crops with medicinal importance. In: Neglected and underutilized crops. CRC Press, pp 27–36. https://doi.org/10.1201/9781003477792-3

25. Sari S, Saylan D (2025) Sustainable strategies for cancer phytomedicine: balancing efficacy and the environment responsibility. In: Sustainable pharmaceutical product development and optimization processes: from eco-design to supply chain integrity. Springer Nature Singapore, Singapore, pp 359–382. https://doi.org/10.1007/978-981-97-9707-3
26. Ahmad Nejhad A, Alizadeh Behbahani B, Hojjati M, Vasiee A, Mehrnia MA (2023) Identification of phytochemical, antioxidant, anticancer and antimicrobial potential of Calotropis procera leaf aqueous extract. Sci Rep 13(1):14716. https://doi.org/10.1038/s41598-023-42086-1
27. Cerri F, Giustra M, Anadol Y et al (2022) Natural products from mangroves: an overview of the anticancer potential of Avicennia marina. Pharmaceutics 14(12):2793. https://doi.org/10.3390/pharmaceutics14122793
28. Pham DC, Shibu MA, Mahalakshmi B et al (2020) Effects of phytochemicals on cellular signaling: reviewing their recent usage approaches. Crit Rev Food Sci Nutr 60(20):3522–3546. https://doi.org/10.1080/10408398.2019.1699014
29. Shoaib S, Islam N, Yusuf N (2022) Phytocompounds from the medicinal and dietary plants: multi-target agents for cervical cancer prevention and therapy. Curr Med Chem 29(26):4481–4506. https://doi.org/10.2174/0929867329666220301114251
30. Soltys L, Olkhovyy O, Tatarchuk T et al (2021) Green synthesis of metal and metal oxide nanoparticles: principles of green chemistry and raw materials. Magnetochemistry 7(11):145. https://doi.org/10.3390/magnetochemistry7110145
31. Radulescu DM, Surdu VA, Ficai A et al (2023) Green synthesis of metal and metal oxide nanoparticles: a review of the principles and biomedical applications. Int J Mol Sci 24(20):15397. https://doi.org/10.3390/ijms242015397
32. Chakraborty A, Haque SM, Dey D et al (2022) Phytogenic silver nanoparticles from tissue-cultured Kaempferia angustifolia—an underutilized medicinal herb: a comparative antibacterial study on urinary pathogens. J Genet Eng Biotechnol 20(1):131. https://doi.org/10.1186/s43141-022-00414-4
33. Suresh D, Nethravathi PC, Rajanaika H et al (2015) Green synthesis of multifunctional zinc oxide (ZnO) nanoparticles using Cassia fistula plant extract and their photodegradative, antioxidant and antibac terial activities. Mater Sci Semicond Process 31:446–454. https://doi.org/10.1016/j.mssp.2014.12.023
34. Venkata ALK, Sivaram S, Srilakshman G et al (2022) Review on terpenoid mediated nanoparticles: significance, mechanism, and biomedical applications. Adv Nat Sci Nanosci Nanotechnol 13(3):033003. https://doi.org/10.1088/2043-6262/ac865f
35. Basha SK, Govindaraju K, Manikandan R et al (2010) Phytochemical mediated gold nanoparticles and their PTP 1B inhibitory activity. Colloids Surf B Biointerfaces 75(2):405–409. https://doi.org/10.1016/j.colsurfb.2009.09.008
36. Su Y, Li Y, Yi Q et al (2025) Insight into the mechanisms and clinical relevance of antifungal heteroresistance. J Fungi 11(2):143. https://doi.org/10.3390/jof11020143
37. Chen X, Shen M, Feng Q et al (2025) Epidemiological characteristics, antifungal susceptibility, and mortality factors of candidemia in adults at a tertiary teaching hospital in Zunyi, China (2016–2023). BMC Infect Dis 25(1):726. https://doi.org/10.1186/s12879-025-11021-2
38. Hetta HF, Melhem T, Aljohani HM et al (2025) Beyond conventional antifungals: combating resistance through novel therapeutic pathways. Pharmaceuticals 18(3):364. https://doi.org/10.3390/ph18030364
39. Dladla M, Gyzenhout M, Marias G et al (2024) Azole resistance in Aspergillus fumigatus-comprehensive review. Arch Microbiol (7):305. https://doi.org/10.1007/s00203-024-04026-z
40. Mroczyńska M, Brillowska-Dąbrowska A (2020) Review on current status of echinocandins use. Antibiotics 9(5):227. https://doi.org/10.3390/antibiotics9050227
41. Ngece K, Ntondini TL, Khwaza V et al (2024) Polyene-based derivatives with antifungal activities. Pharmaceutics 16(8):1065. https://doi.org/10.3390/pharmaceutics16081065

42. Hammoudi Halat D, Younes S, Mourad N et al (2022) Allylamines, benzylamines, and fungal cell permeability: a review of mechanistic effects and usefulness against fungal pathogens. Membranes 12(12):1171. https://doi.org/10.3390/membranes12121171
43. Sigera LSM, Denning DW (2023) Flucytosine and its clinical usage. Ther Adv Infect Dis 10:20499361231161387. https://doi.org/10.1177/20499361231161387
44. Al-Afify S, Adel RA, Arafat EES (2025) Griseofulvin: a comprehensive review of its antifungal, pharmacological, and clinical implications. AlSalam Int J Pharm. https://doi.org/10.21608/aijp.2025.356963.1003
45. Biersack B (2025) Special issue "antifungal drug discovery: progresses, challenges, opportunities". Int J Mol Sci 26(5):2065. https://doi.org/10.3390/ijms26052065
46. Branda F, Petrosillo N, Ceccarelli G et al (2025) Antifungal agents in the 21st century: advances, challenges, and future perspectives. Infect Dis Rep 17(4):91. https://doi.org/10.3390/idr17040091
47. Cui X, Wang L, Lue Y et al (2022) Development and research progress of anti-drug resistant fungal drugs. J Infect Public Health 15(9):986–1000. https://doi.org/10.1016/j.jiph.2022.08.00
48. Singh B, Dahiya M, Kumar V et al (2025) Biofilm and antimicrobial resistance: mechanisms, implications, and emerging solutions. Microbiol Res 16(8):183. https://doi.org/10.3390/microbiolres16080183
49. Boakye YD, Osafo N, Danquah CA et al (2019) Antimicrobial agents: antibacterial agents, anti-biofilm agents, antibacterial natural compounds, and antibacterial chemicals. In: Antimicrobials, antibiotic resistance, antibiofilm strategies and activity methods. IntechOpen. https://doi.org/10.5772/intechopen.82560
50. Salomoni R, Léo P, Montemor AF et al (2017) Antibacterial effect of silver nanoparticles in Pseudomonas aeruginosa. Nanotechnol Sci Appl:115–121. https://doi.org/10.2147/NSA.S133415
51. Shemetov AA, Nabiev I, Sukhanova A (2012) Molecular interaction of proteins and peptides with nanoparticles. ACS Nano 6(6):4585–4602. https://doi.org/10.1021/nn300415x
52. Long N, Li F (2024) Antifungal mechanism of natural products derived from plants: a review. Nat Prod Commun 19(8):1934578X241271747. https://doi.org/10.1177/1934578X241271747
53. Slavin YN, Bach H (2022) Mechanisms of antifungal properties of metal nanoparticles. Nano 12(24):4470. https://doi.org/10.3390/nano12244470
54. Pandey S, Mishra S, Mishra A et al (2025) Recent developments in electron microscopy and applications of open-source electronic hardware in electron microscopes. Integr Ferroelectr 241(7–9):745–762. https://doi.org/10.1080/10584587.2025.2483654
55. Ding J, Yan Z, Peng L et al (2025) Inhibitory effects of berberine on fungal growth, biofilm formation, virulence, and drug resistance as an antifungal drug and adjuvant with prospects for future applications. World J Microbiol Biotechnol 41(1):5. https://doi.org/10.1007/s11274-024-04223-4
56. Tuan DA, Uyen PVN, Masak J (2024) Advances in compounds targeting regulatory mechanisms of biofilm formation in unicellular eukaryotes: insights from 2024 research and applications in Vietnam. https://doi.org/10.20944/preprints202409.1406.v2
57. Ivanov M, Ćirić A, Stojković D (2022) Emerging antifungal targets and strategies. Int J Mol Sci 23(5):2756. https://doi.org/10.3390/ijms23052756
58. Yuan X, Yang F, Wang Y et al (2024) Scopoletin negatively regulates the HOG pathway and exerts antifungal activity against Botrytis cinerea by interfering with infection structures, cell wall, and cell membrane formation. Phytopathol Res 6(1):1. https://doi.org/10.1186/s42483-023-00219-4

Chapter 9
Role of Underutilized Plant-Derived Metal-Based Nanoparticles Against Breast Cancer

Neetika Kimta and Garima Bhanwala

Abstract Human and animal lives are significantly influenced by the kingdom of plants. The second most common cause of cancer-related fatalities in women is breast cancer. It is still difficult to prevent breast cancer worldwide since it is a multi-step process that involves several cell types. With the existing treatments, advanced breast cancer with distant organ metastases is deemed incurable. Green-synthesized nanoparticles appear to be a new tool for medication delivery in tumour cells and predictive biomarkers for cancer diagnosis. Small size active and passive targeting, the capacity to connect multiple targeting moieties, controlled release, and site-specific targeting are only a few of the special features that nanotechnology offers. This chapter emphasizes the emerging significance of underutilized plant groups, specifically pteridophytes, in the green synthesis of metal-based nanoparticles and their application in breast cancer treatment. Through methods including the unique characteristics of nanoparticles themselves as well as the antioxidative and anticancer qualities present in plants, they can either cause damage to tumour cells or protect healthy cells. Therefore, to maintain healthy cells, underutilized medicinal plants should be further investigated in the field of nano-pharmacology to strengthen their particular targeting of the malignant cells.

Keywords Nanoparticles · Green synthesis · Underutilized plants · Mechanistic insights · Bioactive compounds

Abbreviations

DLS	Dynamic Light Scattering
FESEM	Field Emission Scanning Electron Microscopy

N. Kimta (✉) · G. Bhanwala
School of Biological and Environmental Sciences, Shoolini University of Biotechnology and Management Sciences, Solan, India

© The Author(s), under exclusive license to Springer Nature Switzerland AG 2026

E. Nepovimova et al. (eds.), *Underutilized Plants Based-Nanoparticles: Herbal Remedy to Treat Human Diseases*, https://doi.org/10.1007/978-3-032-20384-7_9

FTIR	Fourier-Transform Infrared Spectroscopy
HRTEM	High-Resolution Transmission Electron Microscopy
IC_{50}	Inhibitory Concentration 50
NPs	Nanoparticles
ROS	Reactive Oxygen Species
SEM	Scanning Electron Microscopy
TEM	Transmission Electron Microscopy
XRD	X-Ray Diffraction

9.1 Introduction

Cancer is a condition that arises when there is an uncontrolled growth of normal cells, which later on causes genetic instability and, in the end, the mutations accumulate in both cells and tissues, thus the cells become cancerous [1]. Researchers have made a lot of attempts to solve the problem of cancer, but still, it is very difficult to find a proper cure. Due to the presence of side effects and the development of drug resistance, the efficacy of conventional chemotherapy in achieving complete cancer treatment is significantly compromised [2]. Recent years have witnessed a shift from chemical-based synthetic pharmaceuticals to plant-derived medicines [3]. Formulations from plants have the advantage of being naturally composed; nevertheless, they have less stability and reduced adsorption when in contact with bodily fluids [4]. The use of nanomedicine has been a breakthrough in cancer therapy in the last few years, making it possible to deliver drugs to cancer cells, thus saving healthy tissue [5]. The physicochemical features that make nanoparticles (NPs) capable of anticancer activity can be a combination of inherent properties, e.g. antioxidant action, and properties that are changed by external factors, for instance, hyperthermia induced by infrared rays or magnetic fields [6, 7]. Nanotechnology is a term that is mainly concerned with the science and engineering part of the creation of particles within the range of 1–100 nanometres [8]. The biosynthesis of nanoparticles by means of green technology using biomaterials such as microorganisms and plants is a new, safe, and effective way to produce nanoparticles. Biological methods yield nanoparticles with better size and shape uniformity than conventional synthetic methods. Sustainable methods consume less energy and release eco-friendly products and waste [9]. Green nanotechnology makes it possible for drugs to be released slowly from the carrier, specifically at the diseased area, which subsequently increases therapeutic effect and biocompatibility. Nano formulations, specifically nanoparticles created by integrating specific metals with plant extracts, are currently the focus of research aimed at addressing existing limitations.

Breast cancer has overtaken lung cancer as the most prevalent cancer globally, with 2.26 million cases documented annually [10]. NPs enhance the compatibility and bioavailability of natural bioactives for the treatment of various chronic disorders, including cancer [11]. Numerous researchers regard nanomedicines as the future of medicine, given their application in biological and pharmacological contexts [12]. Angiosperms are the predominant taxonomic group of plants investigated for the green synthesis of NPs. There is a growing body of data on algae, whereas information regarding green synthesis involving pteridophytes, bryophytes, and gymnosperms remains scarce. The cosmopolitan distribution of angiosperms, relative to other plant groups, is a clear factor contributing to their increased involvement in green synthesis [13]. Pteridophytes are less frequently investigated in the field of nanotechnology compared to other plant classes [14]. It represents an intermediary group of plants that connect non-tracheophytes and spermatophytes. Green synthesis of nanoparticles has been largely figured out in angiosperms, but only a handful of studies have been published for different plant groups like bryophytes, gymnosperms, and particularly pteridophytes. The prevalence of angiosperms in nanoparticle research can be accounted for by their global distribution and easy accessibility [13]. Pteridophytes, a bridge group between non-vascular and seed plants, have been under the radar of researchers, though they are packed with phytochemicals and have a historical background of being used in the local healthcare system [14].

A comprehensive review of relevant literature indicates that the green synthesis of nanomaterials from underutilized plants is a viable and environmentally friendly method, which can be further extended in biomedical applications. The present chapter intends to elucidate the novel, eco-friendly, and efficient methods of cancer treatment by emphasizing the exploration of this fascinating research field. The literature review indicates that pteridophytes possess significant potential for the green synthesis of nanomaterials, especially in biomedical applications.

9.2 Underutilized Plants as Sources for Green Nanoparticle Synthesis

Plant-mediated green nanoparticle synthesis is a favoured method due to its typical requirement for neutral pH and the ability to occur at room temperature [15]. Various parts of a plant, i.e. fruits, seeds, stems, peels, leaves, and flowers, have been found to be effective in the synthesis of metal NPs of different shapes and sizes. Plant extracts are loaded with bioactive compounds like alkaloids, phenolic acids, polyphenols, proteins, carbohydrates, and terpenoids, which help in the reduction and stabilization of metal ions [16]. To produce metal nanoparticles using a green chemistry method based on plants, one has to mix a plant extract that will serve as a reducing agent with a metal precursor under suitable conditions [8]. The process is picturized in three phases: (1) the activation one, wherein metal ions are reduced by phytoconstituents thus, nucleation of reduced metal atoms takes place; (2) the growth stage, via which the adhesion of the smaller nanoparticles leads to larger nanoparticles; and (3) finally, the termination stage, wherein nanoparticles obtain their ultimate structure [9].

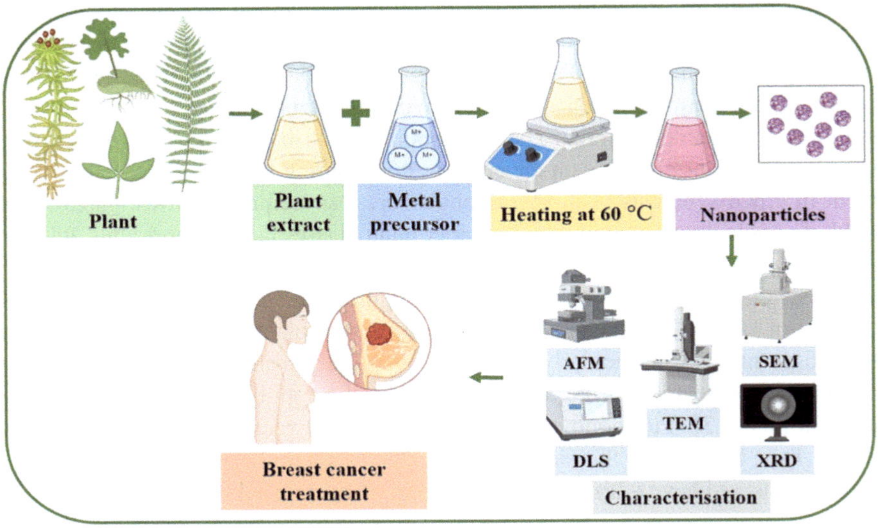

Fig. 9.1 Green synthesis of metallic nanoparticles using plant extract

Typically, nanoparticles are made through various physical and chemical methods. Figure 9.1 is a schematic representation of the fabrication of biogenic nanoparticles from plant sources. The NPs produced have a high surface energy, which, in turn, can cause them to agglomerate. To avoid such agglomeration, a capping agent is generally used [17]. One of the most prominent features of the synthesis of nanoparticles is the alteration in the colour of the reaction mixture during incubation, becoming a more intense colour from the lighter one with time. Phyto-synthesis leads to the generation of nanoparticles, whose characteristics are influenced by the specific extract employed. The domain of green synthesis has predominantly concentrated on investigating various species of *Adiantum* and *Azolla* [18]. Phytosynthesized nanoparticles, closely associated with natural extracts, are employed in applications where these extracts have historically been utilized, including antioxidant and antimicrobial domains [9].

Besides being a potential source for eco-friendly and cheap nanoparticle production, underutilized plants have not been extensively researched for this purpose. For instance, ferns like *Pteris vittata*, which have been ignored as a source for the synthesis of nanomaterials, are full of polyphenols, tannins, and flavonoids [9]. Correspondingly, catechin and epicatechin-containing *N. exaltata* leaf extracts were utilized in the biofabrication of zinc oxide nanoparticles that exhibited considerable cytotoxic activity [19]. Besides that, gymnosperms, being a good source for terpenoids and flavonoids, have been instrumental in the production of silver nanoparticles and are reported to have induced apoptosis in cancer cells [20]. Besides that, *Azolla pinnata* (aquatic plant) is among other underutilized species that possess polysaccharides and alkaloids that would make nanoparticle production easier with the potential of applications in biomedical engineering [21].

The wide array of characterization approaches provides solid evidence for the generation as well as the operable features of the metal NPs obtained from these sources of the residual biomass. Nanoparticle formation is affirmed by UV–Visible spectroscopy from the unique surface plasmon resonance absorption bands. Fourier-transform infrared (FTIR) spectroscopy reveals the functional groups involved in the reduction and stabilization of metal ions [22]. X-ray diffraction (XRD) is used to determine crystalline structure, phase composition, and particle size, along with scanning electron microscopy (SEM) and transmission electron microscopy (TEM), which give the morphological details and size distribution analysis. Dynamic light scattering (DLS) and zeta potential measurements are used to get the hydrodynamic diameter and colloidal stability of the nanoparticles produced, thereby assuring reproducibility and usefulness in a variety of fields [9].

9.3 Metal-Based Nanoparticles in Breast Cancer Therapy

Metallic nanoparticles are a major focus in breast cancer therapy due to the extraordinary manner in which they can elevate targeted drug delivery and, at the same time, provoke cytotoxic effects in breast cancer cells. Such nanoparticles take advantage of the enhanced permeability and retention effect to deliver less and more drugs to be delivered to the tumour tissues, thus increasing drug bioavailability and at the same time decreasing side effects [23]. In terms of their mechanism of action, the metallic NPs produce reactive oxygen species that cause oxidative stress, the sources of which are mitochondria and DNA, and the fragmentation of which leads to cell apoptosis, in breast cancer cells [24]. Besides, some metallic NPs help make tumour cells more vulnerable to radiation or chemo by producing ROS and changing the tumour microenvironment, such as by removing the oxygen deficiency, which is a common cause of resistance to therapy [24]. Metallic NPs can also modulate the immune system by, for example, unleashing the activity of innate immune cells and through the production of immunogenic cell death, which results in the body's immune system fighting cancer [25]. Metallic NPs typically generate ROS, leading to oxidative stress and apoptosis [26]. Apoptosis initiates with the activation of apoptotic proteins, followed by DNA damage, mitochondrial disintegration, apoptosome assembly, and subsequent cell shrinkage [27]. Caspases, a class of protease enzymes, play a crucial role in the apoptotic process [28]. The plant-based green synthesis of metal nanoparticles has garnered significant attention owing to their non-toxic, non-hazardous, cost-effective, and environmentally friendly characteristics [8]. Different metals NPs, via their exceptional physicochemical features, have been very promising for breast cancer diagnosis and therapy. As a result, metallic nanoparticles should be considered as the future agents that combine therapy with diagnostics. These NPs allow for tumour imaging, targeted drug delivery, and the use of hyperthermia to kill cells.

Underutilized medicinal plants have unveiled fresh prospects and opportunities for cancer treatment. They serve as a source of novel chemicals through

Table 9.1 Green-synthesized AgNPs against human breast cancer cell lines

Plant	Plant part	Shape	Size (nm)	Cancer cell line	IC_{50} value	References
Elephantopus scaber	Leaf	Spherical	59	MCF-7	10 μg/ml (GI_{150})	[32]
Fagonia indica	Whole	Spherical	10–60	MCF-7	12.35 μg/ml	[33]
Ganoderma neo-japonicum	Whole	Spherical	5–8	MDA-MB-231	6.0 μg/ml	[26]
Litchi chinensis	Leaf	Spherical	41–55	MCF-7	40.9 μg/ml	[34]
Putranjiva roxburgi	Seed	Spherical	8	MDA-MB-231	0.26 mg/ml	[35]
Solanum trilobatum	Fruit	Spherical, polygonal	41.90	MCF-7	30 μg/ml	[36]

phytochemical screening and facilitate the development of innovative cancer treatment methods, such as the green synthesis of AgNPs [29]. AgNPs obtained by the green synthesis method with the help of the extracts of the underutilized plants have shown to be very effective in the fight against cancer in breast cancer cells. As a matter of fact, the AgNPs produced from such underutilized plants as *Artemisia monosperma* and *Eugenia roxburghii* have led to pronounced cytotoxic effects on breast carcinoma (MCF-7) cells, and cell viability was found to decrease with an increase in the concentration. These green fabricated AgNPs attain excellent anticancer potency even at lower concentrations than those of chemically synthesized ones, which is evidence that natural plant-derived phytochemicals act as a positive factor in the activation of nanoparticles [30, 31]. Currently, AgNPs have been extensively investigated for breast cancer diagnosis and treatment (Table 9.1). Green-synthesized AgNPs, coated with phytochemicals, have enhanced biological activity compared to AgNPs produced using chemical methods. NPs exhibit site-specific and targeted activity, so they can augment therapeutic efficacy as they traverse impermeable membranes and modulate immune system responses, making them suited for cancer treatment.

9.4 Mechanistic Insights

Biogenically synthesized metal nanoparticles from underutilized plants have become a potential tool in breast cancer therapy owing to their distinct physicochemical properties and compatibility with biological systems. How metal nanoparticles fight breast cancer is a question that is being researched, and it is believed to involve multiple pathways [37]. Metal NPs generate ROS, which may lead to oxidative stress and apoptosis in cancer cells [38–40]. Metal NPs might also induce

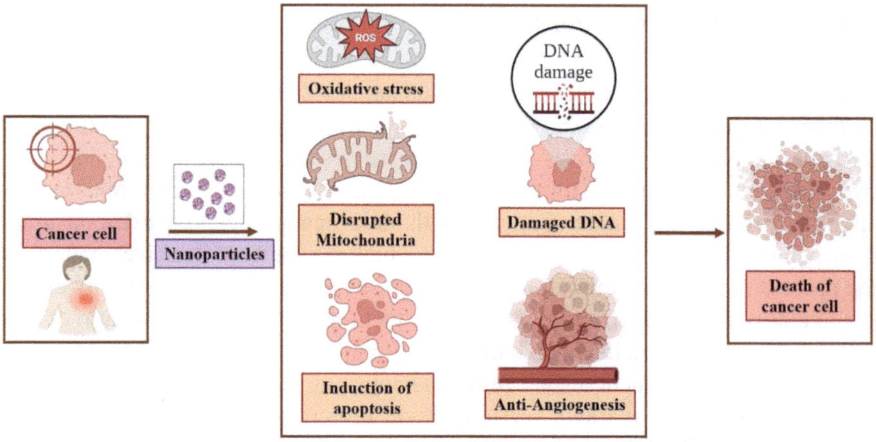

Fig. 9.2 Mechanistic insights into the anticancer activity of metallic nanoparticles against breast cancer cells

apoptosis in cancer cells by the activation of pro-apoptotic signalling pathways [41]. Besides that, metal nanoparticles are reported to have anti-inflammatory effects that may prevent cancer cells' proliferation and metastasis [37, 39, 41]. In addition, they have shown the ability to restrain angiogenesis, the process through which new blood vessels are formed, thus possibly limiting the growth and spread of breast cancer cells (Fig. 9.2) [42].

9.4.1 Oxidative Stress

Metal nanoparticles induce the production of reactive oxygen species in the breast cancer cells through the Fenton reaction. This happens when metal nanoparticles are internalized by cancer cells of the breast and react with cellular iron. Iron interacts with hydrogen peroxide (H_2O_2) coming from the cells to produce hydroxyl radicals (OH^-) [39]. Hydroxyl radicals are very reactive, and thus, they can induce oxidation of the biological constituents that co-exist with them, namely, DNA, proteins, and lipids. The generation of ROS mainly explains the side effects of metal NPs; however, other mechanisms that can be linked to ROS generation also exist. For example, the oxidative damage may finally cause cell death through apoptosis or necrosis [37, 39, 40]. Metal NPs can worsen the situation of mitochondria by inducing oxidative stress. Metal NPs can be the source of ROS that may attack the biological components, such as mitochondria, and thus cause oxidative stress [40]. This may lead the cellular energy production to decrease, thus causing mitochondrial dysfunction, which, in turn, can lead to apoptosis [43].

9.4.2 Induction of Apoptosis

The apoptosis in breast cancer cells mechanism, developed by metal nanoparticles, involves various pathways, and these pathways were divided into two main types: intrinsic and extrinsic apoptosis. Metal nanoparticles are capable of inducing oxidative stress and disrupting mitochondrial integrity; thus, they can release pro-apoptotic proteins and initiate apoptosis [44]. Metal nanoparticles have demonstrated their potential to activate caspase enzymes, which are the most decisive enzymes in the process of apoptosis execution. Metal NPs can damage mitochondria by elevating oxidative stress, interrupting energy production, and changing the ratio of pro- and anti-apoptotic proteins, which eventually will lead to cellular dysfunction and death [39]. The movement of Zn^{2+} ions can happen through ion channels; therefore, the activity of Bcl-2 markers (anti-apoptotic proteins) is inhibited [39]. Green-synthesized AgNPs stimulate the production of reactive oxygen species that eventually lead to cellular death. The formation of reactive oxygen species causes damage to signal transduction pathways, which is followed by cellular apoptosis. The production of hydrogen peroxide causes a change in the mitochondrial membrane potential that results in the unbinding of respiration [43].

9.4.3 DNA Damage

One of the consequences of the overproduction of ROS is the occurrence of oxidative DNA damage and mitotic cell death. This might result in the subsequent activation of autophagic or mitophagic cell death that involves changes in gene expression through changes in transcription factors [39]. Bcl-2 is a protein that plays an essential role in the regulation of programmed cell death in cells. According to the research, inhalation of metallic nanoparticles may have a potential impact on Bcl-2 expression and other genes related to apoptosis. The metal nanoparticles-induced changes in the transcription factors of cancer cells most probably depend on the type of cancer, the dose of nanoparticles, and the exposure conditions. Additional studies are required to identify which transcription factors are influenced by metal nanoparticles and to determine the mechanisms by which they regulate gene expression [41, 44].

9.4.4 Suppress Angiogenesis

The exact mechanism by which metal NPs interfere with breast cancer cells is not completely specified; however, it is assumed that it affects several pathways [42]. Metal NPs have shown the ability to reduce the work of matrix metalloproteinases, enzymes that help to dissolve the extracellular matrix and make angiogenesis easier

[41]. Metal NPs have been reported to exhibit antioxidant properties, which may help in lowering oxidative stress and inflammation in cancer cells, thereby preventing angiogenesis [37, 38]. Metal NPs have demonstrated the ability to inhibit the Akt signalling pathway that is instrumental in the regulation of angiogenesis [45]. Green-synthesized AgNPs, due to their anti-angiogenic potency, may offer a novel avenue for breast cancer treatment. Autophagy-induced cellular breakdown resulting in cell death is suggested as an additional mechanism postulated for the anticancer potential of AgNPs [46].

9.5 Conclusion and Future Perspectives

Breast cancer is becoming more common all around the world; therefore, it's important to look for new, safe, and effective treatment options. Green nanotechnology, particularly the use of nanoparticles derived from metals found in plants, is a sustainable and long-lasting approach. These biogenic nanoparticles combine the healing properties of medicinal plants with the ability of nanocarriers to deliver drugs to specific areas. Because they are biocompatible, can be made in an environmentally benign way, and have the potential for long-term, controlled drug release, they are great prospects for cancer treatment. Pteridophytes and other plant groupings that aren't used enough have a lot of promise for making nanoparticles because they have a lot of phytochemicals and have been used in traditional medicine. Even though they aren't getting much attention in current nanotechnology research, early studies suggest that they could be good biofactories for making effective anticancer nanomaterials. This creates new ways to combine old knowledge with new biomedical advances. We can make next-generation, eco-friendly, and very effective cancer treatments by using both phytochemicals and nanotechnology. For plant-based metal nanoparticles to reach their full potential in clinical cancer treatment, botanists, nanotechnologists, pharmacologists, and oncologists will need to continue collaborating across disciplines.

Acknowledgments We acknowledge the Chancellor and Vice-Chancellor of Shoolini University of Biotechnology and Management Sciences. This research did not receive any specific grant from funding agencies in the public, commercial, or not-for-profit sectors.

Author's Contribution Garima Bhanwala: Writing—original draft, data curation, conceptualization. Neetika Kimta: Writing—original draft, validation, supervision, data curation.

Competing Interest The author declares no competing financial or personal interests.

Ethical Approval This chapter is based on a literature review and secondary data; hence, no ethical approval was required.

References

1. Iqbal J, Abbasi BA, Mahmood T, Kanwal S, Ali B, Shah SA, Khalil AT (2017) Plant-derived anticancer agents: a green anticancer approach. Asian Pac J Trop Biomed 7(12):1129–1150. https://doi.org/10.1016/j.apjtb.2017.10.016
2. Biswas AK, Islam MR, Choudhury ZS, Mostafa A, Kadir MF (2014) Nanotechnology-based approaches in cancer therapeutics. Adv Nat Sci Nanosci Nanotechnol 5(4):043001. https://doi.org/10.1088/2043-6262/5/4/043001
3. Davatgaran-Taghipour Y, Masoomzadeh S, Farzaei MH, Bahramsoltani R, Karimi-Soureh Z, Rahimi R, Abdollahi M (2017) Polyphenol nanoformulations for cancer therapy: experimental evidence and clinical perspective. Int J Nanomedicine 12:2689–2702. https://doi.org/10.2147/IJN.S131973
4. Jeevanandam J, Aing YS, Chan YS, Pan S, Danquah MK (2017) Nanoformulation and application of phytochemicals as antimicrobial agents. In: Antimicrobial nanoarchitectonics. Elsevier, pp 61–82. https://doi.org/10.1016/B978-0-323-52733-0.00003-3
5. Menon S, Ks SD, Kumar V (2018) Selenium nanoparticles: a potent chemotherapeutic agent and an elucidation of its mechanism. Colloids Surf B Biointerfaces 170:280–292. https://doi.org/10.1016/j.colsurfb.2018.06.006
6. Loutfy SA, Al-Ansary NA, Abdel-Ghani NT, Hamed AR, Mohamed MB, Craik JD, Eldin TA, Abdellah AM, Hussein Y, Hasanin MT, Elbehairi SE (2015) Anti-proliferative activities of metallic nanoparticles in an in vitro breast cancer model. Asian Pac J Cancer Prev 16(14):6039–6046
7. Vinardell MP, Mitjans M (2015) Antitumor activities of metal oxide nanoparticles. Nano 5(2):1004–1021. https://doi.org/10.3390/nano5021004
8. Bhanwala G, Kimta N, Ravi K (2025) Nanopesticides in plant protection: potentials, mechanistic insights, and future perspectives against phytopathogens and insect pests. Physiol Mol Plant Pathol:102851. https://doi.org/10.1016/j.pmpp.2025.102851
9. Kimta N, Chauhan A, Puri S, Kumari A, Sharma R, Kumar A, Kapoor D (2025) Phytomediated copper oxide nanoparticles derived from the fronds of *Adiantum venustum* D. Don: evaluation of their biomedical potential. Appl Biochem Biotechnol 197(1):398–426. https://doi.org/10.1007/s42247-024-00715-z
10. Sung H, Ferlay J, Siegel RL, Laversanne M, Soerjomataram I, Jemal A, Bray F (2021) Global cancer statistics 2020: GLOBOCAN estimates of incidence and mortality worldwide for 36 cancers in 185 countries. CA Cancer J Clin 71(3):209–249. https://doi.org/10.3322/caac.21660
11. Al-kawmani AA, Alanazi KM, Farah MA, Ali MA, Hailan WAQ, Al-Hemaid FM (2020) Apoptosis-inducing potential of biosynthesized silver nanoparticles in breast cancer cells. J King Saud Univ Sci 32(4):2480–2488. https://doi.org/10.1016/j.jksus.2020.04.002
12. Nankar RP, Raman M, Doble M (2016) Nanoformulations of polyphenols for prevention and treatment of cardiovascular and metabolic disorders. In: Emulsions. Academic Press, pp 107–151. https://doi.org/10.1016/B978-0-12-804306-6.00004-0
13. Das RK, Brar SK (2013) Plant mediated green synthesis: modified approaches. Nanoscale 5(21):10155–10162. https://doi.org/10.1039/C3NR02548A
14. Fierascu I, Fierascu IC, Brazdis RI, Baroi AM, Fistos T, Fierascu RC (2020) Phytosynthesized metallic nanoparticles—between nanomedicine and toxicology: a brief review of 2019's findings. Materials 13(3):574. https://doi.org/10.3390/ma13030574
15. Velusamy P, Kumar GV, Jeyanthi V, Das J, Pachaiappan R (2016) Bio-inspired green nanoparticles: synthesis, mechanism, and antibacterial application. Toxicol Res 32(2):95–102. https://doi.org/10.5487/TR.2016.32.2.095
16. Marshall AT, Haverkamp RG, Davies CE, Parsons JG, Gardea-Torresdey JL, van Agterveld D (2007) Accumulation of gold nanoparticles in *Brassica juncea*. Int J Phytoremediation 9(3):197–206. https://doi.org/10.1080/15226510701376026

17. Sharma M, Thakur P, Gaur P, Rani GM, Rustagi S, Talreja RK, Chaudhary V (2023) Cancer treatment and toxicity outlook of nanoparticles. Environ Res 237:116870. https://doi.org/10.1016/j.envres.2023.116870
18. Kimta N, Chauhan A, Puri S, Kumari A, Sharma R, Kumar A, Kapoor D (2025) Phytomediated copper oxide nanoparticles derived from the fronds of *Adiantum venustum* D. Don: evaluation of their biomedical potential. Appl Biochem Biotechnol 197(1):398–426. https://doi.org/10.1007/s12010-024-05031-0
19. Aboul-Soud MA, Siddique R, Fozia F, Ullah A, Rashid Y, Ahmad I, Zaghloul NS, Al-Rejaie SS, Mohany M (2023) Antiplatelet, cytotoxic activities and characterization of green-synthesized zinc oxide nanoparticles using aqueous extract of Nephrolepis exaltata. Environ Sci Pollut Res 30(29):73870–73880
20. Kumari R, Saini AK, Kumar A, Saini RV (2020) Apoptosis induction in lung and prostate cancer cells through silver nanoparticles synthesized from Pinus roxburghii bioactive fraction. JBIC J Biol Inorg Chem 25(1):23–37
21. Korbekandi H, Chitsazi MR, Asghari G, Najafi RB, Badii A, Iravani S (2014) Green biosynthesis of silver nanoparticles using Azolla pinnata whole plant hydroalcoholic extract. Green Process Synth 3(5):365–373
22. Kimta N, Chauhan A, Puri S, Kumari A, Sharma R, Kumar A, Kapoor D (2024) Evaluating the biomedical potential of phytomediated silver-doped zinc oxide nanoparticles derived from n-butanol fraction of Adiantum venustum D. Don. J Indian Chem Soc 101(11):101364. https://doi.org/10.1016/j.jics.2024.101364
23. Al-Samydai A, Abu Hajleh MN, Al-Sahlawi F, Nsairat H, Khatib AA, Alqaraleh M, Ibrahim AK (2024) Advancements of metallic nanoparticles: a promising frontier in cancer treatment. Sci Prog 107(4):00368504241274967
24. Alven S, Gandidzanwa S, Ngalo B, Poswayo O, Madanhire T, Aderibigbe BA, Tshentu Z (2024) Platinum group metals nanoparticles in breast cancer therapy. Pharmaceutics 16(9):1162
25. Zarenezhad E, Kanaan MHG, Abdollah SS, Vakil MK, Marzi M, Mazarzaei A, Ghasemian A (2023) Metallic nanoparticles and their potential role in breast cancer immunotherapy via trained immunity provocation. Biomedicine 11(5):1245
26. Ullah I, Khalil AT, Ali M, Iqbal J, Ali W, Alarifi S, Shinwari ZK (2020) Green-synthesized silver nanoparticles induced apoptotic cell death in MCF-7 breast cancer cells by generating reactive oxygen species and activating caspase-3 and -9 enzyme activities. Oxidative Med Cell Longev 2020:1215395. https://doi.org/10.1155/2020/1215395
27. Jain N, Jain P, Rajput D, Patil UK (2021) Green synthesized plant-based silver nanoparticles: therapeutic prospective for anticancer and antiviral activity. Micro Nano Syst Lett 9(1):5. https://doi.org/10.1186/s40486-021-00131-6
28. Li Y, Ke Y, Zou H, Wang K, Huang S, Rengarajan T, Wang L (2019) Gold nanoparticles synthesized from Strychni semen and their anticancer activity in cholangiocarcinoma cell (KMCH-1). Artif Cells Nanomed Biotechnol 47(1):1610–1616. https://doi.org/10.1080/21691401.2019.1594860
29. Klefenz H (2004) Nanobiotechnology: from molecules to systems. Eng Life Sci 4(3):211–218. https://doi.org/10.1002/elsc.200402090
30. Al-Sheddi ES, Alsohaibani N, bin Rshoud N, Al-Oqail MM, Al-Massarani SM, Farshori NN, Malik T, Al-Khedhairy AA, Siddiqui MA (2023) Anticancer efficacy of green synthesized silver nanoparticles from Artemisia monosperma against human breast cancer cells. S Afr J Bot 160:123–131
31. Khuda F, Gul M, Ali Khan Khalil A, Ali S, Ullah N, Shafiq Khan M, Nazir S, Irum Khan S, Mehtap Büyüker S, Almawash S, Shafique M (2023) Biosynthesized silver nanoparticles using Alnus nitida leaf extract as a potential antioxidant and anticancer agent. ACS Omega 8(33):30221–30230
32. Poudel BK, Soe ZC, Ruttala HB, Gupta B, Ramasamy T, Thapa RK, Gautam M, Ou W, Nguyen HT, Jeong JH, Jin SG (2018) In situ fabrication of mesoporous silica-coated silver–gold hollow nanoshell for remotely controllable chemo-photothermal therapy via phase-

change molecule as gatekeepers. Int J Pharm 548(1):92–103. https://doi.org/10.1016/j.ijpharm.2018.06.056
33. Iqbal MJ, Ali S, Rashid U, Kamran M, Malik MF, Sughra K, Zeeshan N, Afroz A, Saleem J, Saghir M (2018) Biosynthesis of silver nanoparticles from leaf extract of Litchi chinensis and its dynamic biological impact on microbial cells and human cancer cell lines. Cell Mol Biol 64(13):42–47. https://doi.org/10.14715/cmb/2018.64.13.9
34. Shinde AS, Mendhulkar VD (2020) Antiproliferative activity of Elephantopus scaber-mediated silver nanoparticles against MCF-7, A-549, SCC-40 and COLO-205 human cancer cell lines. Asian J Pharm Clin Res 13(2):163–167. https://doi.org/10.22159/ajpcr.2020.v13i2.36497
35. Gurunathan S, Raman J, Malek SNA, John PA, Vikineswary S (2013) Green synthesis of silver nanoparticles using Ganoderma neo-japonicum Imazeki: a potential cytotoxic agent against breast cancer cells. Int J Nanomedicine 8:4399–4413. https://doi.org/10.2147/IJN.S51881
36. El-Naggar NEA, Hussein MH, El-Sawah AA (2017) Bio-fabrication of silver nanoparticles by phycocyanin: characterization, in vitro anticancer activity against breast cancer cell line, and in vivo cytotoxicity. Sci Rep 7(1):10844
37. Wang J, Lee JS, Kim D, Zhu L (2017) Exploration of zinc oxide nanoparticles as a multitarget and multifunctional anticancer nanomedicine. ACS Appl Mater Interfaces 9(46):39971–39984. https://doi.org/10.1021/acsami.7b11219
38. Adeyemi JO, Onwudiwe DC, Oyedeji AO (2022) Biogenic synthesis of CuO, ZnO, and CuO–ZnO nanoparticles using leaf extracts of Dovyalis caffra and their biological properties. Molecules 27(10):3206. https://doi.org/10.3390/molecules27103206
39. Ansari A, Ali A, Khan N, Umar MS, Owais M (2022) Synthesis of steroidal dihydropyrazole derivatives using green ZnO nanoparticles and evaluation of their anticancer and antioxidant activity. Steroids 188:109113. https://doi.org/10.1016/j.steroids.2022.109113
40. George BP, Rajendran NK, Houreld NN, Abrahamse H (2022) *Rubus*-capped zinc oxide nanoparticles induce apoptosis in MCF-7 breast cancer cells. Molecules 27(20):6862. https://doi.org/10.3390/molecules27206862
41. Aalami AH, Mesgari M, Sahebkar A (2020) Synthesis and characterization of green zinc oxide nanoparticles with antiproliferative effects through apoptosis induction and microRNA modulation in breast cancer cells. Bioinorg Chem Appl 2020:8817110. https://doi.org/10.1155/2020/8817110
42. Pouresmaeil V, Haghighi S, Raeisalsadati AS, Neamati A, Homayouni-Tabrizi M (2021) Anti-breast cancer effects of green-synthesized zinc oxide nanoparticles using carob extracts. Anti Cancer Agents Med Chem 21(3):316–326. https://doi.org/10.2174/1871520620666200721132522
43. Avalos A, Haza AI, Mateo D, Morales P (2016) Interactions of manufactured silver nanoparticles of different sizes with normal human dermal fibroblasts. Int Wound J 13(1):101–109. https://doi.org/10.1111/iwj.12244
44. Chen D, Li B, Lei T, Na D, Nie M, Yang Y, Xie C, He Z, Wang J (2021) Selective mediation of ovarian cancer SKOV3 cell death by pristine carbon quantum dots/Cu_2O composite through targeting matrix metalloproteinases, angiogenic cytokines and cytoskeleton. J Nanobiotechnol 19(1):68. https://doi.org/10.1186/s12951-021-00813-8
45. Song WJ, Jeong MS, Choi DM, Kim KN, Wie MB (2019) Zinc oxide nanoparticles induce autophagy and apoptosis via oxidative injury and pro-inflammatory cytokines in primary astrocyte cultures. Nano 9(7):1043. https://doi.org/10.3390/nano9071043
46. Balkrishna A, Sharma VK, Das SK, Mishra N, Bisht L, Joshi A, Sharma N (2020) Characterization and anticancer effect of Putranjiva roxburghii seed extract–mediated silver nanoparticles on human colon (HCT-116), pancreatic (PANC-1) and breast (MDA-MB-231) cancer cell lines: a comparative study. Int J Nanomedicine 15:573–585

Chapter 10
Role of Underutilized Plant-Derived Metal Oxide-Based Nanoparticles Against Prostate Cancer

Nidhi Bhardwaj, Indu Bhardwaj, Kanika Dulta, Eugenie Nepovimova, Kamil Kuca, Monika Choudhary, Parteek Mandyal, Ambika Choudhary, and Lukas Peter

Abstract Prostate cancer (PC) is the second most often detected cancer among men and the fifth leading cause of cancer-related deaths worldwide. In its early stages, PC is typically asymptomatic and progresses slowly, often requiring active surveillance. While no definitive protective measures have been established, certain lifestyle changes, such as increased consumption of green vegetables, reduced intake of high-fat foods, and regular physical activity, may help to lower the risk. Current satisfying approaches, predominantly allopathic (medications, radiations,

N. Bhardwaj
Center of Advanced Innovation Technologies, VŠB-Technical University of Ostrava, Ostrava-Poruba, Czech Republic

Department of Chemistry, Faculty of Science, University of Hradec Kralove, Hradec Kralove, Czech Republic

I. Bhardwaj
School of Health Sciences, Department of MLT, Abhilashi University, Mandi, India

K. Dulta · L. Peter
Center of Advanced Innovation Technologies, VŠB-Technical University of Ostrava, Ostrava-Poruba, Czech Republic

E. Nepovimova (✉)
Department of Chemistry, Faculty of Science, University of Hradec Kralove, Hradec Kralove, Czech Republic
e-mail: eugenie.nepovimova@uhk.cz

K. Kuca
Center of Basic and Applied Research, Faculty of Informatics and Management, University of Hradec Kralove, Hradec Kralove, Czech Republic

M. Choudhary · A. Choudhary
School of Biological and Environmental Sciences, Shoolini University of Biotechnology and Management Sciences, Solan, Himachal Pradesh, India

P. Mandyal
School of Advanced Chemical Sciences, Shoolini University, Solan, Himachal Pradesh, India

© The Author(s), under exclusive license to Springer Nature Switzerland AG 2026

E. Nepovimova et al. (eds.), *Underutilized Plants Based-Nanoparticles: Herbal Remedy to Treat Human Diseases*, https://doi.org/10.1007/978-3-032-20384-7_10

and surgery), are associated with several adverse effects, including loss of muscle mass, reduced libido, and bone thinning. In light of these boundaries, research has begun to shift toward green therapeutics, utilizing plant-derived compounds for more holistic treatment strategies. However, some plant-based bioactive compounds can inadvertently induce oxidative stress by unsettling the body's antioxidant protective system. To address this challenge, nanotechnology has developed as a promising solution. Through the improvement of bioavailability, biodegradability, half-life, and controlled release, along with protection from enzymic hydrolysis and minimizing dose-dependent toxicity reasonably, the combination of plant materials and nanotechnology surpasses the shortfalls of both plant extracts and traditional nanoparticle fabrication processes. Therefore, this chapter discovers the potential of underutilized plants and their derived metal oxide nanoparticles in the treatment of PC.

Keywords Nanoparticles · Underutilized plants · Prostate cancer · Medicine · Mechanism

Abbreviations

ADT	Androgen Deprivation Therapy
Akt/PI3K	Protein kinase B/phosphatidylinositol 3-kinase
AR	Androgen receptors
ATM	Ataxia-Telangiectasia Mutated
BRCA1	Breast cancer gene 1
BRCA2	Breast cancer gene 2
DNA	Deoxyribose nucleic acid
MAPK/ERK	Mitogen-activated protein kinase/extracellular signal-regulated kinase
MRP	Multidrug resistance associated protein
NF-κB	Necrosis factor kappa B
PSA	Prostate-specific antigen
STAT	Signal transducer and activator of transcription

10.1 Introduction

From the perspective of both epidemiology and genetics, prostate cancer (PC) is a multilayered disease that usually targets middle-aged men (45 and 60 years). Nowadays, it is already the leading cause of cancer-related mortality in Western countries [1]. It has a high prevalence in developed countries; actually, GLOBOCAN reported about 1,276,106 new cases in 2018, leading to approximately 358,989 deaths globally [2]. In fact, the prevalence of PC varies with ethnic group and geographic location worldwide. For instance, the world's highest reported incidence rate is among Afro-American men [3]. Beyond these demographic features, however, the clinical heterogeneity of PC mirrors molecular heterogeneity and lineage plasticity, in which neoplastic cells, upon receiving environmental stimuli, switch between different lineages and phenotypic states [4]. Prostate cancer can also be classified, according to its response to testosterone stimulation, as either androgen-sensitive or androgen-insensitive, which directly impacts available therapies [5].

10.1.1 Pathophysiology

(a) *Androgen Receptors (AR)*

AR is a ligand-dependent transcription factor (TF) belonging to the nuclear receptor family that is activated by many androgens, like dihydrotestosterone (DHT), testosterone, and other androgenic steroids [6, 7]. Following its ligand binding, AR dimerizes and is phosphorylated before translocation into the nucleus. In the ligand's absence, it is found in the cytoplasm and bound to some chaperone proteins. In the nucleus, the AR-ligand dimer binds to 'androgen response elements (AREs)' within the promoter region of androgen-responsive genes [8, 9]. This recruitment mediates the association with several co-regulator epigenetic factors and proteins, thus forming a transcriptionally active complex that initiates expression of these genes downstream [6, 7]. This subsequent gene expression activates growth and survival of PC cells [10] (Fig. 10.1).

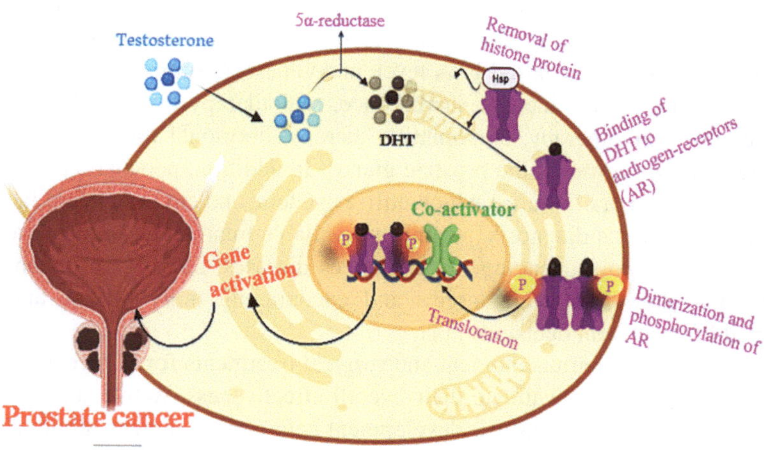

Fig. 10.1 The mechanism of AR activation is triggered by binding the dihydrotestosterone, followed by dimerization of the receptors, phosphorylation (P), and finally its translocation to the nucleus. Within the nucleus, the receptors bind to respective DNA sequences, bring about co-activators, and cause gene transcription, which results in enhanced cell proliferation and promotion of cancer

(b) *Gene Mutations*
From a genetic perspective, somatic copy number alterations (SCNAs) and single nucleotide polymorphisms (SNPs) are two examples of the mutations that alter DNA sequences and cause cancer [11]. Men with PC have identifiable mutations/alterations in the genes responsible for DNA repair, including BRCA2, ATM, and BRCA1 [12]. These mutations cause excessive division of cells and can be inherited by an individual or passed down from generation to generation. Apart from DNA repair genes, other genes that function as PC biomarkers include HOX genes, RNase L (HPC1, lq22), ELAC2/HPC2 (17p11), and others [13].

(c) *Epigenetic Changes*
DNA methylation changes, chromatin protein expression, and histone modifications are important epigenetic changes that can be utilized in the detection and classification of PC. Since DNA methylation is reversible, it is also a potential target for epigenetic cancer therapy. Such methylation inhibits gene expression, thus the gene can no longer produce its corresponding RNA or protein. A recent bioinformatic study has found 37 microRNAs (miRNAs) with CpG islands in their upstream parts of the stem-loop structures of their precursor. Unmethylated CpG islands play a vital role in ensuring the correct functioning of the cells, but in the case of prostate cancer, they are usually abnormally methylated, leading to epigenetic silencing of the associated miRNAs [14, 15].

10.1.2 Current Treatment and Its Disadvantages

In PC, the prognostic factors that are used to determine the best course of treatment for PC include age, comorbidities, baseline urinary function, clinical tumor, node, metastasis (TNM) stage, initial PSA level, and Gleason's score [16]. The common treatments used for PC include surveillance, radiotherapy (radium-223 therapy), cryotherapy, chemotherapy, combination therapy, hormonal therapy, and radical prostatectomy (removal of the prostate gland). In addition, personalized medicine (PM) is another area of interest which utilizes genetic and environmental factors to determine a patient's diagnosis, prognosis, available treatments, and accurate dosage. Using genome sequencing, it categorizes diseases, identifying patients with tumors that show actionable targets and encouraging more precise and well-informed PC treatment decisions [17].

Though several pharmacological and surgical treatments for PC are available, a number of limitations invalidate their overall effectiveness. Active surveillance is replete with the risk of metastasis development before the treatment, along with tension and the hassle of ongoing clinical follow-up in the patient [18]. Surgical interventions, as potent in their ability to resect tumors, are often accompanied by decreased chances of preserving sexual function and other quality-of-life disabilities. Pharmacotherapies, particularly ADT, are specifically faced with significant challenges owing to adaptive modulations of key signaling pathways such as Akt/

PI3K, MAPK/ERK, AR signaling, and EAR pathways. These molecular adaptations form the basis of drug resistance, limiting the long-term efficacy of hormonal therapies and complicating the management of the disease [19]. Other factors contributing to drug resistance in PC include MRP transporters [20]. For example, it has been reported that the expression of MRP2 is most potent when it comes to resistance to natural product agents, whereas MRP4 expression is in charge of resistance to many chemotherapeutic agents such as methotrexate, cyclophosphamide, nucleoside analogues, and camptothecins [21].

10.1.3 Underutilized Plants and Green Nanotechnology for PC Treatment

According to data on plants, between 40 and 1 lakh species have been utilized for fiber, food, culture, commerce, and medicine over the years [22]. However, the left plant diversity is referred to as "underutilized," and the rural population's economic growth and food security are entirely dependent on those few species of plants. Underutilized wild edible plants (UWEP) are acknowledged as orphaned and neglected species of nature that have been used for centuries for fiber, food, medicine, oil, and other uses. However, because of their short shelf life, limited supply, reputational issues, and low consumer awareness, their significance and use have significantly decreased. Furthermore, their gene pools have suffered greatly as a result of modern agricultural practices, making them "lost" or "neglected" crops [23]. According to the recent findings, several researchers have investigated, verified, and scientifically assessed almost 100 UWEP plant species, particularly food herbs and wild woody plants in Pakistan, northwest Ethiopia, and West Africa [24–26]. These include plant species that are rich in bioactive compounds, such as *Urtica dioica*, *Artemisia scoparia*, *Amaranthus viridis*, *Adhatoda vasica*, and *Galium aparine*. However, following a survey of documented traditional knowledge (TK) of these species in Kenya's western and Nyaza provinces, the spider plant, *Cleome gynamandra*, was recognized as a UWEP species as well [27]. Similarly, the wild weed *Indigofera glandulosa* can be found in Indonesia, India, and North Australia. Ghane et al. [28] even claim that the last-mentioned plant is rich in vitamins and minerals and that it could be used as a vegetable.

Additionally, because of their high dietary fiber content, these underutilized plants have a variety of pharmacological actions. For example, the wild species *Cyperus esculentus*, also known as tigernut, is used to treat gastrointestinal disorders, diabetes, obesity, colon cancer, and coronary heart disease [24]. Following multiple trials of anti-epileptic medications, *Casimiroa edulus*, a UWEF species of Ethiopia, demonstrated an exceptional source of anti-epileptic compounds [29]. *C. esculentus* contains high-quality oil and essential amino acids that improve adult health. *Allium chinense*, *Dioscorea hamiltonii*, *Zingiber officinale* Roscoe, and *Colocasia esculenta* (L.) roots and tubers have been recommended for the treatment of anemia, piles, wounds, cancer, and skin eruptions [30]. In addition to these species, *Citrullus lanatus*, *Ceratotheca sesamoides*, *Cajanus cajan*, *Pennisetum glaucum*, *Sphenostylis stenocarpa*, *Vigna subterranean*, *Cleome gynandra*, *Sesamum indicum*, and *Corchorus olitorius* contain a lot of micronutrients and antioxidants [24] and demonstrate a variety of pharmacological functions. For instance, the roots, leaves, and seeds of these plant species can be used to treat liver problems, burns, injuries, infections, hypertension, breast cancer, and sexually transmitted diseases [31].

Despite numerous advantages of these plant species for human health, certain disadvantages associated with their function retrieve their superior position from the treatment track. For example, in complex mixtures, these plant phytocompounds show synergistic effects, but they sometimes lose their activity due to their isolation, chemical instability, poor bioavailability, solubility, poor brain penetration, insufficient intravenous dosing, lower plasma levels, and lessened biological action under normal conditions [32–34]. To overcome this issue associated with the application, researchers from diverse fields, such as material sciences and nanoscience, are now using plant-derived nano-formulations or green nanotechnology [35].

One of the primary areas of research now is green nanotechnology, which includes the synthesis/manufacturing of nanoparticles (NPs) using biological materials. NPs, which are thought of as the base of nanotechnology, are observed in sizes from 1 to 100 nm. Many types of NPs, such as metal (M) or metal oxides (MO) like copper, silver, gold, and zinc, have garnered a lot of consideration lately due to their fundamental and technological relevance [36]. In fact, a wide range of industries, including biology, biotechnology, medicine, both food and textile, agriculture, and wastewater treatment, use these green nanomaterials [37]. In addition, green nanotechnology offers the material in a safe manner, avoids the use of hazardous materials, concentrates on reducing environmental pollution, and places the onus of product responsibility on manufacturers and scientists [38]. These plant materials aid in NPs reduction, capping, and surface area and morphology protection [39]. Therefore, information regarding the role of metal oxides derived from underutilized plants against PC may be targeted in the current chapter.

10.2 Types of Underutilized Plant-Derived Metal Oxide-Based Nanoparticles Used Against PC

10.2.1 Zinc Oxide NPs (ZnO NPs)

ZnO NPs' green synthesis has drawn more attention in the literature due to their capability to intermingle well with cancer cells. According to the literature [40], certain compounds can bind ZnO NPs in the green process without changing their structure, but they can also introduce specific characteristics that allow the addition of extra tools to combat cancer cells, including prostate cancer. Indeed, the main mechanism by which ZnO NPs attack PC cells is apoptosis, which can be brought

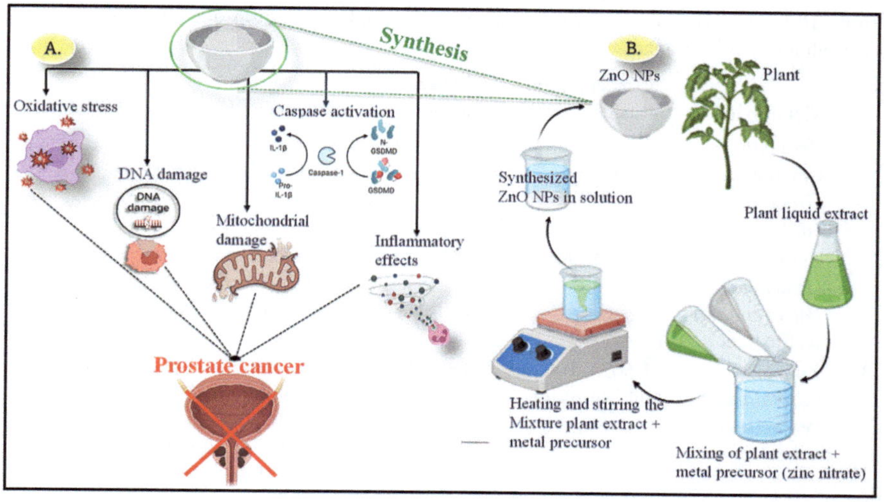

Fig. 10.2 Picture illustrating how ZnO NPs combat PC cells (**a**), as well as ZnO NPs´ synthesis process (**b**). Synthetic process begins with the collection of plant material and the preparation of liquid extract. The extract is then combined with a zinc precursor solution to form ZnO NPs. The prepared mixture is then heated and stirred on a magnetic stirrer hot plate for several hours (variable time depending on experimental conditions). Later, pellets formed at the beaker bottom are harvested, washed successively by centrifugation, dried by a muffle furnace or oven, and finally crushed to obtain ZnO NPs in powder form

on by oxidative stress, caspase activation, DNA damage, mitochondrial damage, and pro-inflammatory effects [40] (Fig. 10.2; Table 10.1 [information on different metal-oxides derived from underutilized plants, presented separately from the main text]). The most important underutilized plant species used for the manufacturing of zinc oxide NPs are listed below:

(a) *Crotalaria verrucosa* L.

Blue rattle pod (*C. verrucosa*), a low shrub distributed in all India from the "Himalayas to Ceylon" [41], has been used for ages in traditional medicine for treating numerous diseases, including leprosy, biliousness, rheumatism, skin allergy, diarrhea, dyspepsia, fever, and dysentery. Phytochemical analysis revealed that the plant contains numerous bioactive constituents—such as glycosides, flavonoids, tannins, steroids, alkaloids, and polyphenolic compounds—which exhibit high cytotoxicity against cancer cells and hence justify its potential use in anticancer treatment [42]. Sana et al. [43] synthesized ZnO NPs using the plant's leaf extract and then tested them using the MTT assay in in vitro cell lines, including DU145 and HeLa cell lines (2 mg/mL). The authors showed that prepared NPs displayed an IC_{50} value of 6.30 μg/mL for DU145 and 7.07 μg/mL against HeLa cell line at a concentration of 100 μg/mL, respectively. In contrast, the positive control (leaf extract) showed an IC_{50} of 15.72 μg/mL and 22.30 μg/mL, respectively, against the same cell lines.

Table 10.1 Summary of metal oxide nanoparticles synthesized from underutilized plants and their inhibitory effects against PC

Sr. No.	Metal oxide	Concentration used	Model name	Positive outcome	References
1.	ZnO NPs (microalgae *Gracilaria edulis*)	10–100 μg/mL; 48 h	PC3 cell	ZnO NPs exhibited 28.55 μg/mL IC_{50} against cell line	[83]
2.	ZnO NPs (*Leucaena leucocephala*) leaf extract	62.5–1000 μg/mL	PC3 and MCF-7 cells	NPs marched 85% "cell death in PC3 cells, and 75% cell death in MCF-7 at 1 mg/mL"	[84]
3.	ZnO NPs (*Crotalaria verrucosa* leaf extract)	1.5625–100 μg/mL; overnight	DU-145, and HeLa cell line	ZnO NPs exhibited the highest inhibition against DU-145 (IC50: 6.30 μg/mL), while 7.07 μg/mL (IC50) for HeLa cell lines at 100 μg/mL	[43]
4.	ZnO NPs (fig leaves)	0, 20, 40, 80, 160, 320 ppm; 24 and 48 h	PC3 cell line	Small size of ZnO NPs, i.e., 32.55 nm, exhibited a great cytotoxic effect marked by increased apoptosis and decreased viability	[85]
5.	SeO_2 NPs (*Deverra tortuosa*)	1.56, 3.125, 6.25, 12.5, 25, 50, and 100 μg/mL	PC3 cell line	SeO_2 NPs showed 69.89% inhibition on PC3 cells at 100 μg/mL	[58]
6.	Fe_2O_3 NPs *Phyllostachys heterocycla*	0.01, 0.1, 1, 10, 100 μg/mL; 48 h	PC-3 cell line	Fe_2O_3 NPs exhibited IC50 ranging from 2.31 to 14.9 μg/mL. Author stated that the integration of humic acid with NPs enhances their cytotoxic potential	[86]
7.	FeONPs *Ajuga macrosperma*	100, 200, 400, 600, 800 μg/mL; 48 h	PC3 cell line	FeONPs exhibited 2.65 μg/mL IC50, and also showed antibacterial (*Staphylococcus aureus*, *E. coli*, *Bacillus subtilis*, and *Pseudomonas aeruginosa*) potential	[87]

(b) *Hyssopus officinalis* L.

H. officinalis is an endemic Iranian species, which is mostly used for medicinal and traditional purposes. It is used as a food flavoring agent despite having a bitter taste and is known for its strong antibacterial and antioxidant effects [43, 44]. Rahimi Kalateh Shah Mohammad et al. [45] prepared zinc oxide NPs using plant leaf extract and evaluated the anti-cancer activity in vitro against PC3 cells and in vivo using "Balb/C mice." The green-synthesized zinc oxide NPs were very cytotoxic (IC_{50}: 8.07 μg/mL [24 h] and 5 μg/mL [48 h]). In addition, apoptosis effects in PC3 cells were dose-dependent with 26.6%, 44%, and 80% apoptotic cells being detected for NPs concentrations of 4.88 μg/mL, 9.76, and 19.53 μg/mL, respectively, compared to untreated control cells using "acridine orange–propidium iodide (AO-PI) staining." Furthermore, ZnO NPs also demonstrated impaired spermatogenesis, sperm motility, and sperm maturation process.

(c) *Ruellia tuberosa* L.
Often referred to as the "cracker plant," R. tuberosa is a significant ethno-medicinal plant that is found throughout Central America, Africa, South, and Southeast Asia. Various plant parts have archaeologically been utilized to cure a wide range of illnesses like lung disorders, kidney-related issues, and "sexually transmitted diseases (STDs)" [46]. The literature review also dictates that this plant has anti-pyretic, anti-hypertensive, anti-diabetic, antimicrobial, antioxidant, anti-inflammatory, analgesic, and anti-cancer properties [47–49]. In 2025, Guo et al. [50] synthesized zinc oxide NPs using leaf extract and tested their cytotoxic potential. The authors selected a range of concentrations (5, 10, 15, 20, 25, 30, 35, 40, 45, 50 μg/mL) to test against DU145 and normal prostate epithelial cells (HPE-15). Based on their outcomes, NPs exhibited minimum toxicity towards HPE-15 cells and high cytotoxicity towards DU145 cells (IC50: 26.82 μg/mL) in comparison with the untreated cells (control), and positive control called Docetaxel (DTX). According to the Western blot data, ZnO NPs downregulated STAT3 and upregulated p53 expression, suggesting that they have potential against PC through the STAT3 and p53 signaling pathways.

(d) *Peganum harmala* L.
P. harmala is a herb that falls in the Zygophyllaceae family and is native to arid regions from the east Mediterranean to northern India and Iran [51]. Traditionally, this plant has been used to control a variety of conditions, including gastrointestinal disorders, Parkinson's disease, cardiovascular disorders, and various types of cancer [52]. In 2023, Shabestarian [53] synthesized zinc oxide NPs using the seed extract of this plant and tested them for PC treatment using three different cell lines: HepG2, A2780, and PC3. To test diverse concentrations of ZnO NPs (3.9, 7.8, 15.6, 31.2, and 62.5 μg/mL), the authors used the MTT assay. According to their findings, NPs displayed diverse cytotoxicity profiles against cancer cell lines. For example, ZnO NPs displayed 47.36 μg/mL IC50 value against PC3, 49.96 μg/mL against HepG2, and 11.65 μg/mL against A2780, respectively.

(e) *Deverra tortuosa*
The wild perennial bushy plant, *D. tortuosa*, generally grows in stony and sandy plains [54]. Bioactive compounds such as essential oils, glycosides, flavonoids, terpenoids, unsaturated sterols, and furanocoumarins have been found in different parts of this plant by Abdelwahed et al. [55] and Abdallah et al. [56]. El-Alfi et al. [57] listed several ethno-pharmacological uses of the plant as an analgesic, carminative, diuretic, remedy for intestinal parasites, stomach pain, anti-asthmatic, and scorpion sting. El-Amier et al. [58] synthesized zinc oxide NPs from methanolic extract and evaluated the cytotoxic activity of these bio-synthetic NPs against different cell lines like HePG2, PC3, MCF-7, and WI-38 cells. Various concentration

ranges from 1.56 to 100 μg/mL were applied, and IC_{50} values of 97.25 μg/mL were recorded for WI-38, 25.28 μg/mL for HePG2, 44.53 μg/mL for MCF-7, and lastly 56.90 μg/mL for PC3 were recorded. These authors stated that the harmfulness of these NPs may be caused by their aggregated structure, spherical shape, and nanometer size (<100 nm).

10.2.2 Copper Oxide Nanoparticles or CuO NPs

Copper oxide NPs have attracted great attention for their multidirectional applications in medicine and industry. Aside from their extensively documented antimicrobial efficacy against a wide array of bacteria and microorganisms [59], CuO NPs are also highly anti-cancerous in nature. There have been reports that document their ability to inhibit cancer cell growth by inducing cytotoxicity [60]. Hanagata et al. [61] demonstrated that copper ions leached from copper oxide NPs resulted in the "arrest of the cell cycle, DNA damage, and inhibition of proliferating cell nuclear antigen (PCNA) in lung epithelial A549 cells. Similarly, CuO NPs were found to induce apoptosis mechanisms in PC cells" [62], and further showed that they attain value as therapeutic agents in cancer treatment.

(a) *Aesculus hippocastanum* L.
A. hippocastanum, buckeyes or horse chestnut [63], is a fast-growing ornamental tree used for treating various diseases and belongs to the family Hippocastanaceae [64]. Scientifically, this plant contains various bioactive compounds that play a vital role in treating diabetes, inflammation, and gastroprotection [65]. This plant species, according to traditional knowledge, is utilized for vascular and peripheral diseases like hematoma, varicose veins, and venous disorders [66], pain, cardiac venous insufficiency [67], diarrhea, fever, abrasion of the epidermis, and enlargement of the prostate gland [68]. Aksel et al. [62] utilized this plant for the synthesis of CuO NPs and then conducted various tests on these CuO NPs for their potential to act against prostate cancer. The cytotoxicity test was performed using an androgen-independent PC3 cell line along with variable concentrations of CuO NPs (5, 10, 20, 40 μM; 24 h). According to the obtained results, concentrations above 10 μM demonstrated cytotoxicity to PC3 cell lines of about 50%.

10.2.3 Hematite (α-Fe_2O_3)

Fe_2O_3 is a chemically and thermodynamically stable nanoparticle with a densely packed oxygen lattice and a central hexagonal, rhombohedral geometry [69]. In addition to its applications in sensing, data storage, catalysis, pigments, soil remediation, cosmetics, water treatment, and photochemical cells, it also showed promise as a drug delivery system, cell conjugates, imaging, and reference materials for assessing the characteristics of other metallic oxides due to their lower toxicity and biocompatibility [70]. Fe_2O_3 protects the cells from cancer by preventing necrosis factor kappa B from moving from the cytoplasm to the nucleus. Indeed, NF-κB causes the activation of multiple genes that are entailed in angiogenesis, anti-apoptosis, metastasis, proliferation, and other processes. Fe_2O_3 inhibits NF-κB by keeping membrane receptors bound, which further prevents the receptors from binding to the stress signal. According to Naz et al. [71], this process prevents NF-κB from freely translocating into the nucleus and instead keeps it bound to its inhibitor, IκB.

(a) *Rhus punjabensis*

The indigenous people of Pakistan make extensive use of the roughly 250 species of flowering plants that make up the genus *Rhus*. This genus is a member of the *Anacardiaceae* family, which primarily comprises shrubs and trees with simple or compound (unisexual/bisexual) flowers, alternating leaves, and resinous bark [72]. As members of this genus, *R. punjabensis* is often used in traditional treatments to cure a variety of illnesses, including hepatitis, conjunctivitis, diuresis, ulcer wound healing, diarrhea, leucorrhea, and dysentery [73]. In another study of Tabassum et al., the antioxidant, types of phytochemical, antileishmanial, anticancer, antimicrobial, and protein kinase inhibitory properties of *R. punjabensis* have been also examined [74]. Regarding PC, Naz et al. [71] synthesized metallic oxide nanoparticles (Fe_2O_3 NPs) and tested their anticancer potential against the PC3 and DU-145 cell lines. By using the sulforhodamine B (SRB) cytotoxicity assay, the authors found that Fe_2O_3 NPs demonstrated significant activity against the DU-145 (IC_{50}: 12.79 μg/mL) and PC3 cell lines (IC_{50}: 45.2 μg/mL).

10.3 Other Nanoparticles Derived from Underutilized Plants for PC Treatment

(a) *Leucas aspera*

The Lamiaceae family includes the ayurvedic plant *L. aspera*, which possesses many therapeutic properties. Known by the common name "Thumbai," this plant is grown all over India, from the "Himalayas to Ceylon." This plant has long been used to treat snake bites, chronic skin eruptions, and chronic rheumatism, as well as being used as an insecticide [75]. Alkaloids, phenols, flavonoids, tannins, and glycosides abound in the plant, which confer it with antifungal, cytotoxic, antioxidant, and antimicrobial properties [76]. Mohan et al. [77] synthesized *L. aspera* NPs using the whole plant and further tested their cytotoxic potential against PC3 cell lines using the MTT assay. In this assay, the authors prepared three different concentrations of NPs as follows: 0.0625, 0.125, and 0.25 mg/mL for 24 and 48 h. Their results demonstrated that nanoparticles prepared from *L. aspera* showed cytotoxic effects in both concentration + time dependent manner against PC3 cells. Furthermore, cellular uptake investigation, in vitro hemolysis assay, cell migration assay, and cell aggregation studies recognized their anti-cancerous activity.

(b) *Baliospermum montanum (Willd.) Muell Arg*

A part of the Euphorbiaceae family, *B. montanum* is a woody medicinal plant [78]. This plant has been used historically to treat several properties, including immunomodulatory, anticancer, antimicrobial, anthelmintic, and anti-tumor [79]. Additionally, according to preliminary phytochemical analysis, this plant's anticancer and antioxidant qualities are credited to the existence of tannins, triterpenoids, flavonoids, and glycosides [80]. Moreover, this plant has also been used historically to treat abdominal tumors [81]. Cherian et al. [82] synthesized stable *B. montanum* nanoparticles in 2015 using ethanol and aqueous extract. Thereafter, they tested them towards PC3 cell line. The authors selected 2, 1.5, and 1 mg/mL concentrations for NPs made from aqueous extract of the plant and 1.5 mg/mL and 0.5 mg/mL concentrations for nanoparticles made from ethanol extract for the 48-h MTT assay. According to their findings, NPs synthesized from ethanolic extract exhibited more significant cytotoxicity against PC3 cell lines than NPs made from aqueous extract. They proposed that the anticancer properties of ethanolic extracts could be justified by the presence of anticancer compounds (soluble preferably in ethanol) such as triterpenoids, esters, and steroids.

10.4 Conclusion

The first part of the present chapter contains diverse data collected in association with the PC, which mainly comprised triggering causes such as increased binding affinity of androgen receptors to DHT, gene mutations, and epigenetic changes, and therapeutic approaches such as radiotherapy, cryotherapy, chemotherapy, combination therapy, hormonal therapy, and their corresponding disadvantages are also recorded. Along with that, after literature search, it was observed that ZnO NPs held the top position compared to CuO NPs, Fe_2O_3, and FeONPs in PC treatment. They have been mainly derived from *Crotalaria verrucosa*, *Deverra tortuosa*, *Peganum harmala* L, *Ruellia tuberosa*, *Hyssopus officinalis* L. These NPs regulate the PC cells' proliferation by triggering oxidative stress, activation of caspases, DNA injury, mitochondrial injury, and anti-inflammatory action. Another interesting observation revealed the Fe_2O_3 NPs' ability to inhibit NF-κB via holding membrane receptors bound and also blocking the receptors from binding the stress signal. This alteration then precludes the activation of stress-related genes and control proliferation of tumor cells. In addition to all these metal oxide NPs, there are some non-specified NPs in the literature that modulate PC progression such as *Baliospermum montanum* (Willd.) *Muell Arg* and *Leucas aspera*. Based on the studies listed above, the plants that are currently underutilized could be utilized in the form of MONPs, in particular the ZnO NPs form. However, further in vivo and clinical trials are unconditionally needed to ultimately conclude the above-mentioned in vitro findings and offer under-explored plants and their NPs to another future.

Author's Contribution NB—Writing original draft, data collection; KD, IB, AC, MC—Data interpretation, conceptualization; KK, EN, LP—Final formatting, final approval; PM—Data collection.

Declaration of Competing Interests The contributing authors have no conflicts of interest to declare that are applicable to the content of this chapter.

Ethical Approval Not applicable.

References

1. Chen J, Zhang D, Yan W et al (2013) Translational bioinformatics for diagnostic and prognostic prediction of prostate cancer in the next-generation sequencing era. Biomed Res Int 2013(1):901578. https://doi.org/10.1155/2013/901578
2. Rawla P (2019) Epidemiology of prostate cancer. World J Oncol 10(2):63. https://doi.org/10.14740/wjon1191
3. Barbieri CE, Bangma CH, Bjartell A et al (2013) The mutational landscape of prostate cancer. Eur Urol 64(4):567–576. https://doi.org/10.1016/j.eururo.2013.05.029
4. Ziaran S, Varchulova Novakova Z, Bohmer D et al (2015) Biomarkers for determination prostate cancer: implication for diagnosis and prognosis. Neoplasma 62(5):683–691. https://doi.org/10.4149/neo_2015_082
5. Takayama KI (2019) Splicing factors have an essential role in prostate cancer progression and androgen receptor signaling. Biomolecules 9(4):131. https://doi.org/10.3390/biom9040131
6. Dai C, Heemers H, Sharifi N (2017) Androgen signaling in prostate cancer. Cold Spring Harb Perspect Med 7(9):a030452. https://doi.org/10.1101/cshperspect.a030452
7. Yap TA, Smith AD, Ferraldeschi R et al (2016) Drug discovery in advanced prostate cancer: translating biology into therapy. Nat Rev Drug Discov 15(10):699–718. https://doi.org/10.1038/nrd.2016.120
8. Nadal M, Prekovic S, Gallastegui N (2017) Structure of the homodimeric androgen receptor ligand-binding domain. Nat Commun 8(1):14388. https://doi.org/10.1038/ncomms14388
9. van Royen ME, van Cappellen WA, de Vos C et al (2012) Stepwise androgen receptor dimerization. J Cell Sci 125(8):1970–1979. https://doi.org/10.1242/jcs.096792
10. Davis ID (2022) Combination therapy in metastatic hormone-sensitive prostate cancer: is three a crowd? Ther Adv Med Oncol 14:17588359221086827. https://doi.org/10.1177/17588359221086827
11. Moran A, O'hara C, Khan S et al (2012) Risk of cancer other than breast or ovarian in individuals with BRCA1 and BRCA2 mutations. Fam Cancer 11(2):235–242. https://doi.org/10.1007/s10689-011-9506-2
12. Adhyam M, Gupta AK (2012) A review on the clinical utility of PSA in cancer prostate. Indian J Surg Oncol 3(2):120–129. https://doi.org/10.1007/s13193-012-0142-6
13. Sekhoacha M, Riet K, Motloung P et al (2022) Prostate cancer review: genetics, diagnosis, treatment options, and alternative approaches. Molecules 27(17):5730. https://doi.org/10.3390/molecules27175730
14. Cornblatt G, Yegnasubramanian S, Lee B et al (2008) Epigenetic regulation of microRNAs in prostate cancer. Cancer Res 68(9_Supplement):52–52
15. Enikeeva K, Rafikova G, Sharifyanova Y et al (2024) Epigenetics as a key factor in prostate cancer. Adv Biol 8(5):2300520
16. Trewartha D, Carter K (2013) Advances in prostate cancer treatment. Nat Rev Drug Discov 12(11):823. https://doi.org/10.1038/nrd4068

17. Mateo L, Duran-Frigola M, Gris-Oliver A et al (2020) Personalized cancer therapy prioritization based on driver alteration co-occurrence patterns. Genome Med 12(1):78. https://doi.org/10.1186/s13073-020-00774-x
18. Costello AJ (2020) Considering the role of radical prostatectomy in 21st century prostate cancer care. Nat Rev Urol 17(3):177–188. https://doi.org/10.1038/s41585-020-0287-y
19. Sekhoacha M, Riet K, Motloung P (2022) Prostate cancer review: genetics, diagnosis, treatment options, and alternative approaches. Molecules 27(17):5730. https://doi.org/10.3390/molecules27175730
20. Obligacion R, Murray M, Ramzan I (2006) Drug-metabolizing enzymes and transporters: expression in the human prostate and roles in prostate drug disposition. J Androl 27(2):138–150. https://doi.org/10.2164/jandrol.05113
21. Hwang C (2012) Overcoming docetaxel resistance in prostate cancer: a perspective review. Ther Adv Med Oncol 4(6):329–340. https://doi.org/10.1177/1758834012449685
22. Magbagbeola JAO, Adetoso JA, Owolabi OA (2010) Neglected and underutilized species (NUS): a panacea for community focused development to poverty alleviation/poverty reduction in Nigeria. J Econ Int Finance 2(10):208–211
23. Williams JT (2002) Global research on underutilized crops: an assessment of current activities and proposals for enhanced cooperation. Biodiversity International
24. Dansi A, Vodouhè R, Azokpota P et al (2012) Diversity of the neglected and underutilized crop species of importance in Benin. Sci World J 2012(1):932947. https://doi.org/10.1100/2012/932947
25. Osewa SO, Alamu O, Adetiloye IS, Olubiyi MR, Abidogun EA (2013) Use of some neglected and underutilized plant species among rural dwellers in Akinyele Local Government area of Oyo State. GJAS 3(12):817–822. https://doi.org/10.15580/GJAS.2013.12.060913821
26. Tebkew M, Asfaw Z, Zewudie S (2014) Underutilized wild edible plants in the Chilga District, northwestern Ethiopia: focus on wild woody plants. Agric & Food Secur 3:12. https://doi.org/10.1186/2048-7010-3-12
27. Onyango CM, Kunyanga CN, Ontita EG et al (2013) Current status on production and utilization of spider plant (*Cleome gynandra* L.) an underutilized leafy vegetable in Kenya. Genet Resour Crop Evol 60(7):2183–2189. https://doi.org/10.1007/s10722-013-0036-7
28. Ghane SG, Lokhande VH, Ahire ML et al (2010) *Indigofera glandulosa* Wendl.(Barbada) a potential source of nutritious food: underutilized and neglected legume in India. Genet Resour Crop Evol 57(1):147–153. https://doi.org/10.1007/s10722-009-9496-1
29. Satheesh N (2015) Review on distribution, nutritional and medicinal values of *Casimiroa edulus* Llave-an underutilized fruit in Ethiopia. Am-Euras J Agric Environ Sci 15(8):1574–1583. https://doi.org/10.5829/idosi.aejaes.2015.15.8.9585
30. Gill I, Valivety R (1997) Polyunsaturated fatty acids, part 1: occurrence, biological activities and applications. Trends Bitechnol 15(10):401–409
31. Aniya Y, Koyama T, Miyagi C et al (2002) Free radical scavenging and hepatoprotective actions of *Crassocephalum crepidioides* from Okinawa Islands. Jpn J Pharmacol 88:187P–187P
32. Savjani KT, Gajjar AK, Savjani JK (2012) Drug solubility: importance and enhancement techniques. Int Sch Res 2012(1):195727. https://doi.org/10.5402/2012/195727
33. Thilakarathna SH, Rupasinghe HV (2013) Flavonoid bioavailability and attempts for bioavailability enhancement. Nutrients 5(9):3367–3387. https://doi.org/10.3390/nu5093367
34. Luca SV, Macovei I, Bujor A et al (2020) Bioactivity of dietary polyphenols: the role of metabolites. Crit Rev Food Sci Nutr 60(4):626–659. https://doi.org/10.1080/10408398.2018.1546669
35. Sercombe L, Veerati T, Moheimani F et al (2015) Advances and challenges of liposome assisted drug delivery. Front Pharmacol 6:286. https://doi.org/10.3389/fphar.2015.00286
36. Abdelghany TM, Al-Rajhi AM, Yahya R et al (2023) Phytofabrication of zinc oxide nanoparticles with advanced characterization and its antioxidant, anticancer, and antimicrobial activity against pathogenic microorganisms. Biomass Convers Biorefin 13(1):417–430. https://doi.org/10.1007/s13399-022-03412-1

37. Salem SS (2023) A mini review on green nanotechnology and its development in biological effects. Arch Microbiol 205(4):128. https://doi.org/10.1007/s00203-023-03467-2
38. Soni RA, Rizwan MA, Singh S (2022) Opportunities and potential of green chemistry in nanotechnology. Nanotechnol Environ Eng 7(3):661–673. https://doi.org/10.1007/s41204-022-00233-5
39. Bukhari A, Ijaz I, Gilani E et al (2021) Green synthesis of metal and metal oxide nanoparticles using different plants' parts for antimicrobial activity and anticancer activity: a review article. Coatings 11(11):1374. https://www.mdpi.com/2079-6412/11/11/1374#
40. Rodríguez-Barajas N, Ponce-Regalado MD, Segura-Almendárez MS et al (2024) Plant-mediated synthesis and interaction of ZnO against breast and prostate cancer. Results Chem 9:101654. https://doi.org/10.1016/j.rechem.2024.101654
41. Gamble JS (1928) Flora of the presidency of Madras. West, Newman and Adlard
42. Manokari M, Shekhawat MS (2016) Role of *Crotalaria verrucosa* L. extracts in synthesis of zinc oxide nanoparticles. World Sci Res 3(53):230–240
43. Sana SS, Kumbhakar DV, Pasha A et al (2020) *Crotalaria verrucosa* leaf extract mediated synthesis of zinc oxide nanoparticles: assessment of antimicrobial and anticancer activity. Molecules 25(21):4896. https://doi.org/10.3390/molecules25214896
44. Khazaie HR, Nadjafi F, Bannayan M (2008) Effect of irrigation frequency and planting density on herbage biomass and oil production of thyme (*Thymus vulgaris*) and hyssop (Hyssopus officinalis). Ind Crop Prod 27(3):315–321. https://doi.org/10.1016/j.indcrop.2007.11.007
45. Rahimi Kalateh Shah Mohammad G, Karimi E, Oskoueian E et al (2020) Anticancer properties of green-synthesised zinc oxide nanoparticles using *Hyssopus officinalis* extract on prostate carcinoma cells and its effects on testicular damage and spermatogenesis in Balb/C mice. Andrologia 52(1):e13450. https://doi.org/10.1111/and.13450
46. Roosdiana A, Permata FS, Fitriani RI et al (2020) *Ruellia tuberosa* L. extract improves histopathology and lowers malondialdehyde levels and TNF alpha expression in the kidney of Streptozotocin-induced diabetic rats. Vet Med Int *2020*(1):8812758. https://doi.org/10.1155/2020/8812758
47. Chothani DL, Patel MB, Mishra SH et al (2010) Review on *Ruellia tuberosa* (cracker plant). Pharmacogn J 2(12):506–512. https://doi.org/10.1016/S0975-3575(10)80040-9
48. Seerangaraj V, Sathiyavimal S, Shankar SN et al (2021) Cytotoxic effects of silver nanoparticles on *Ruellia tuberosa*: photocatalytic degradation properties. J Environ Chem Eng 9(2):105088. https://doi.org/10.1016/j.jece.2021.105088
49. Singh S, Neeraj S, Deenanath J (2023) A critical review on *Ruellia Tuberosa*, Saccharum Benghalense and Dichanthium Annulatum with huge availability in 'middle ganga segment'. J Pharm Negative Results 14(1):48–754
50. Guo X, Qu H, Lin X et al (2025) Fabrication of zinc oxide nanoparticles using *Ruellia tuberosa* leaf extract induces apoptosis through P53 and STAT3 signalling pathways in prostate cancer cells. Open Life Sci 20(1):20251121. https://doi.org/10.1515/biol-2025-1121
51. Branch S (2012) Etymological review on chemical and pharmaceutical substances of the oriental origin. Int J Anim Veter Adv 4:40–44
52. Moloudizargari M, Mikaili P, Aghajanshakeri S et al (2013) Pharmacological and therapeutic effects of Peganum harmala and its main alkaloids. Pharmacogn Rev 7(14):199. https://doi.org/10.4103/0973-7847.120524
53. Shabestarian H, Tabrizi MH, Movahedi M et al (2023) Green synthesis of Ag-NPs as a metal nanoparticle and ZnO-NPs as a metal oxide nanoparticle: evaluation of the in vitro cytotoxicity, anti-oxidant, anti-angiogenic activities. Nanomed J 10(3):245–258. https://doi.org/10.22038/NMJ.2023.71700.1770
54. Slima DF, Turki ZA, HAA A et al (2021) Structural adaptation of Deverra tortuosa (Desf.) DC. to its natural habitats in Egypt. Egypt J Bot 61(3):781–794. https://doi.org/10.21608/ejbo.2021.41821.1554

55. Abdelwahed A, Hayder N, Kilani S et al (2006) Chemical composition and antimicrobial activity of essential oils from Tunisian *Pituranthos tortuosus* (Coss.) Maire. Flavour Fragr J 21(1):129–133. https://doi.org/10.1002/ffj.1542
56. Abdallah HM, Ezzat SM (2011) Effect of the method of preparation on the composition and cytotoxic activity of the essential oil of *Pituranthos tortuosus*. Z Naturforsch C 66(3–4):143–148. https://doi.org/10.1515/znc-2011-3-408
57. El-Alfi T, Azzazy MF, Assi NS (2019) Pharmacognostical and DNA fingerprint studies on Deverra tortuosa (DESF.) DC. from Sadat area. JESR 9(2 a):52–61. https://doi.org/10.21608/jesr.2019.56089
58. El-Amier YA, Abduljabbar BT, El-Zayat MM et al (2024) Biosynthesis of metal/metal oxide nanoparticles via *Deverra tortuosa*: characterization, GC/MS profiles, and biological potential. Sci Rep 14(1):23522. https://doi.org/10.1038/s41598-024-74471-9
59. Das D, Nath BC, Phukon P et al (2013) Synthesis and evaluation of antioxidant and antibacterial behavior of CuO nanoparticles. Colloid Surf B Biointerfaces 101:430–433. https://doi.org/10.5681/apb.2011.009
60. Sun T, Yan Y, Zhao Y et al (2012) Copper oxide nanoparticles induce autophagic cell death in A549 cells. PLoS One 7:e43442. https://doi.org/10.1371/journal.pone.0043442
61. Hanagata N, Zhuang F, Connolly et al (2011) Molecular responses of human lung epithelial cells to the toxicity of copper oxide nanoparticles inferred from whole genome expression analysis. ACS Nano 5(12):9326–9338. https://doi.org/10.1021/nn202966
62. Aksel M, Karaağaç D, Kaya T et al (2023) Effect of copper oxide nanoparticles on prostate cancer PC3 cells. IJPAS 9(2):344–351. https://doi.org/10.29132/ijpas.1304265
63. Kapusta I, Janda B, Szajwaj B et al (2007) Flavonoids in horse chestnut (Aesculus hippocastanum) seeds and powdered waste water byproducts. J Agric Food Chem 55(21):8485–8490. https://doi.org/10.1021/jf071709t
64. Ćalić D, Bohanec B, Devrnja N et al (2013) Impact of abscisic acid in overcoming the problem of albinism in horse chestnut androgenic embryos. Trees 27(3):755–762. https://doi.org/10.1007/s00468-012-0830-4
65. Woo CSJ, Lau JSH, El-Nezami H (2012) Herbal medicine: toxicity and recent trends in assessing their potential toxic effects. Adv Bot Res 62:365–384. https://doi.org/10.1016/B978-0-12-394591-4.00009-X
66. Kim SE, Kim TH, Park SA et al (2017) Efficacy of horse chestnut leaf extract ALH-L1005 as a matrix metalloproteinase inhibitor in ligature-induced periodontitis in canine model. J Vet Sci 18(2):245–251. https://doi.org/10.4142/jvs.2017.18.2.245
67. Kędzierski B, Kukula-Koch W, Widelski J et al (2016) Impact of harvest time of Aesculus hippocastanum seeds on the composition, antioxidant capacity and total phenolic content. Ind Crop Prod 86:68–72. https://doi.org/10.1016/j.indcrop.2016.03.034
68. Dudek-Makuch M, Matławska IRENA (2011) Flavonoids from the flowers of Aesculus hippocastanum. Acta Pol Pharm 68(3):403–408
69. Shen S, Zhou J, Dong CL et al (2014) Surface engineered doping of hematite nanorod arrays for improved photoelectrochemical water splitting. Sci Rep 4(1):6627. https://doi.org/10.1038/srep06627
70. Ali A, Pan M, Tilly TB et al (2018) Performance of silver, zinc, and iron nanoparticles-doped cotton filters against airborne *E. coli* to minimize bioaerosol exposure. Air Qual Atmos Health 11(10):1233–1242. https://doi.org/10.1007/s11869-018-0622-0
71. Naz S, Islam M, Tabassum S et al (2019) Green synthesis of hematite (α-Fe2O3) nanoparticles using *Rhus punjabensis* extract and their biomedical prospect in pathogenic diseases and cancer. J Mol Struct 1185:1–7. https://doi.org/10.1016/j.molstruc.2019.02.088
72. Kossah R, Nsabimana C, Zhang H et al (2010) Optimization of extraction of polyphenols from Syrian sumac (Rhus coriaria L.) and Chinese sumac (Rhus typhina L.) fruits. Res J Phytochem 4(3):146–153
73. Shabbir A (2012) Rhus coriaria Linn, a plant of medicinal, nutritional and industrial importance: a review. J Anim Plant Sci 22(2):505–512

74. Tabassum S, Ahmed M, Mirza B et al (2017) Appraisal of phytochemical and in vitro biological attributes of an unexplored folklore: Rhus Punjabensis Stewart. BMC Complement Altern Med 17(1):146. https://doi.org/10.1186/s12906-017-1659-6
75. Srinivasan R, Ravali B, Suvarchala P et al (2011) Leucas aspera-medicinal plant: a review. IJPBS 2(2):159
76. Kurup LB, Latha MS (2013) Chemopreventive potential of methanolic extract of Leucas aspera against N-nitrosodiethyl amine (NDEA) induced hepatotoxicity in rats. Int J Agric Environ Biotechnol 6(3):807–814
77. Mohan A, Nair SV, Lakshmanan VK (2017) Leucas aspera nanomedicine shows superior toxicity and cell migration retarded in prostate cancer cells. Appl Biochem Biotechnol 181(4):1388–1400. https://doi.org/10.1007/s12010-016-2291-5
78. Venkatesh P, Mukherjee PK, Pal BC (2011) Acridanone alkaloid in Baliospermum montanum–evaluation of its effect against anaphylaxis. Planta Med 77(17):1947–1949. https://doi.org/10.1055/s-0030-1271198
79. Johnson M, Wesely EG, Hussain MZ et al (2010) In vivo and in vitro phytochemical and antibacterial efficacy of Baliospermum montanum (Willd.) Muell. Arg. Asian Pac J Trop Med 3(11):894–897. https://doi.org/10.1016/S1995-7645(10)60215-5
80. Sulaiman CT, Balachandran I (2012) Total phenolics and total flavonoids in selected Indian medicinal plants. Indian J Pharm Sci 74(3):258. https://doi.org/10.4103/0250-474X.106069
81. Nandagaon VS, Kulkarni AR (2013) In vitro antioxidant and cytotoxicity activity of Bacopa monnieri and baliospermum montanum muell arg. Int J Pharm Appl 4:63–67
82. Cherian AM, Snima KS, Kamath CR et al (2015) Effect of *Baliospermum montanum* nanomedicine apoptosis induction and anti-migration of prostate cancer cells. Biomed Pharmacother 71:201–209
83. Priyadharshini RI, Prasannaraj G, Geetha N et al (2014) Microwave-mediated extracellular synthesis of metallic silver and zinc oxide nanoparticles using macro-algae (Gracilaria edulis) extracts and its anticancer activity against human PC3 cell lines. Appl Biochem Biotechnol 174(8):2777–2790. https://doi.org/10.1007/s12010-014-1225-3
84. Kanagamani K, Muthukrishnan P, Saravanakumar K et al (2019) Photocatalytic degradation of environmental perilous gentian violet dye using leucaena-mediated zinc oxide nanoparticle and its anticancer activity. Rare Met 38(4):277–286. https://doi.org/10.1007/s12598-018-1189-5
85. Saleh WM, Ghdeeb NJ, Kazzaz FF et al (2024) Impact of green synthesis of ZnO nanoparticles using fig leaves on Saos-2, SK-OV3 and PC3 tumor cell line. Appl Nanosci 14(6):827–833. https://doi.org/10.1007/s13204-024-03051-0
86. Abdelhameed RF, Nafie MS, Ibrahim AK et al (2024) Two metallic nanoparticles formulas of Phyllostachys heterocycla extract exhibited potent cytotoxicity against ovarian cancer cells through apoptosis induction. J Clust Sci 35(7):2557–2569. https://doi.org/10.1007/s10876-024-02645-6
87. Parveen S, Gupta V, Kandwal A et al (2023) In vitro antimicrobial and anticancer activity of metallic nanoparticles (Ag and FeO) against human pathogenic bacteria and cancer cell lines. Eng Proc 56(1):294. https://doi.org/10.3390/ASEC2023-15910

Chapter 11
Role of Underutilized Plant-Derived Metal and Metal Oxide-Based Nanoparticles Against Diabetes

Habiba Iqbal, Ayushi Bhaduri, Kanika Dulta, Nitin, and Ajinkya Nene

Abstract Diabetes mellitus is rapidly rising worldwide, increasing almost twice over the last decades and burdening low- and middle-income countries. Rather than reducing hyperglycemia, the current approaches to the management of diabetes do not target the areas of oxidative stress, chronic inflammation, and progressive development of β-cell dysfunction (reduced insulin production by pancreatic β-cells). This calls for new, low-cost and sustainable interventions. Nanotechnology has the ability to provide fine control of drug delivery and solubility, and provide real-time glucose sensing. But traditional nanoparticle synthesis is frequently associated with the usage of toxic reagents and procedures that are expensive and prohibitively costly, bringing into question its safety and scalability. Medicinal plants have a high bioactive phytochemical content, such as flavonoids, alkaloids, terpenoids, and phenolics, and are in general underutilized candidates that can serve as a sound environmentally friendly alternative in the synthesis of nanoparticles. Green nanotechnology employs the plant components as the reducing and stabilizing agents in order to construct metal and metal oxide nanoparticles which incorporate phytochemical antidiabetic effects with the benefits of nanocarrier delivery systems. There is

H. Iqbal
Centre of Excellence in Molecular Biology, University of the Punjab, Lahore, Pakistan

A. Bhaduri
Department of Food Engineering and Technology, Institute of Chemical Technology, Mumbai, India

K. Dulta
Center of Advanced Innovation Technologies, VSB-Technical University of Ostrava, Ostrava-Poruba, Czech Republic

Nitin
School of Applied and Life Sciences, Uttaranchal University, Dehradun, India

A. Nene (✉)
Department of Mechanics and Aerospace Engineering, Southern University of Science and Technology, Shenzhen, China
e-mail: neneajinkya@gmail.com

© The Author(s), under exclusive license to Springer Nature Switzerland AG 2026

E. Nepovimova et al. (eds.), *Underutilized Plants Based-Nanoparticles: Herbal Remedy to Treat Human Diseases*, https://doi.org/10.1007/978-3-032-20384-7_11

emerging evidence that plant-mediated nanoparticles have glucose uptake improvement, insulin sensitivity increase, reduced oxidative stress, as well as controlled and higher bioavailability.

Keywords Diabetes mellitus · Nanotechnology · Underutilized medicinal plants · Green synthesis · Phytochemicals

Abbreviations

CAT	Catalase
DPPH	2,2-diphenyl-1-picrylhydrazyl
GPX	Glutathione peroxidase
IL-1β	Interleukin-1 beta
IL-6	Interleukin-6
MO-NPs	Metal and metal-oxide nanoparticles
MDA	Malondialdehyde
ROS	Reactive oxygen species
SOD	Superoxide dismutase
STZ	Streptozotocin
TNF-α	Tumor necrosis factor-alpha

11.1 Introduction

Diabetes mellitus is a complex, chronic metabolic disorder characterized by sustained hyperglycemia resulting from disturbances in insulin secretion, insulin action, or both. Insulin, produced by pancreatic β-cells located in the islets of Langerhans, plays a central role in regulating glucose homeostasis by promoting glucose uptake in skeletal muscle and adipose tissue while inhibiting hepatic gluconeogenesis and glycogenolysis. This highly organized system is dysfunctional in patients with diabetes [1]. Type 1 diabetes is caused by the autoimmune destruction of β-cells that result in total insulin deficiency. Contrarily, type 2 diabetes, which is the most common, is predetermined with a complex of peripheral insulin resistance, dysfunction of insulin signaling pathways, and progressive dysfunction of β-cells, which leads to insufficient secretion of insulin in comparison to the needs of the metabolism. Due to insulin insufficiency or inefficiency, glucose enters the blood, causing chronic hyperglycemia [2]. The typical symptoms of diabetes are polyuria, polydipsia, polyphagia, fatigue, blurred vision, frequent infections, and accidental weight loss. The metabolic imbalance does not only stop at glucose imbalance, but also lipid abnormalities, oxidative stress, and low-grade chronic inflammation, which enhance tissue damage. In the long run, sustained hyperglycemia is an etiological factor of both microvascular complications (retinopathy, nephropathy, neuropathy) and macrovascular complications (coronary artery disease, stroke, peripheral arterial disease) [3]. Such complications cause a high morbidity, mortality, and economic burden, so diabetes is not simply an endocrine complication, but it is also a major burden on global health issues.

In 2024, there were 589 million adults aged 20–79 years with diabetes, and that is 11.2% of the global adult population [4]. This is estimated to increase to 853 million by 2050, meaning an increase of 46%. Diabetes patients lost 3.4 million lives in the year 2024 alone which translates to someone dying every nine seconds [5]. Diabetes has also been rising tremendously all over the globe with a near to 14% prevalence by 2022, especially in the low and middle-income countries. In 2022, almost 450 million adults aged 30 years or older, approximately 59% of all individuals with diabetes, were not receiving any form of treatment which is an indication of health disparities [6]. The complications of the disease are more than just hyperglycemia as it is a condition that will cause the vascular and non-vascular comorbidities that include cardiovascular disease, kidney failures, retinopathy, neuropathy, and amputations. The prevalence rate of diabetes among adults in the United States has been 15.8% in 2021–2023, whereby 11.3% and 4.5% of adults had diabetes diagnosed and undiagnosed, respectively [7]. According to the Global Burden of Disease study by 2010, the number of disability-adjusted life years (DALYs) caused by diabetes is increasing, and this fact further demonstrates its growing influence in the health and development agendas worldwide.

Regardless of progress in diabetes management, available therapeutic options are still lacking. Oral glucose-lowering medications, including metformin and sulfonylureas, dipeptidyl peptidase 4 (DPP-4) inhibitors, and sodium-glucose cotransporter-2 (SGLT2) inhibitors, predominantly lower blood sugar levels, but do not correct the underlying pathophysiology [8]. Metformin is associated with frequent gastrointestinal effects but infrequent lactic acidosis, whereas sulfonylureas are associated with risks of weight gain and hypoglycemia. Insulin medication is effective, but it is essential to administer it in the long-term, monitored, and with poor compliance. More recently developed agents such as glucagon-like peptide-1 (GLP-1) receptor agonists and SGLT2 inhibitors have cardiovascular and renal protective effects, yet can be cost-prohibitive in lower-income economies [9]. The therapies do not help with oxidative stress, chronic inflammation, and progressive β-cell dysfunction, resulting in long-term complications.

Nanotechnology is a relatively new innovation in diabetes management, as it provides a new approach to both therapy and diagnosis. Nanoscale materials (1–100 nm) enhance drug solubility, surface area, and controlled release, which allow precise delivery and enhance the bioavailability of a therapeutic agent [10]. Furthermore, nanotechnology has facilitated the use of nanosensors and nanobiosensors to monitor glucose through its concentration in real-time, which is more sensitive and convenient for patients. These developments favor stricter glycemic control and better clinical outcomes. In spite of these advantages, the traditional production of nanoparticles, such as physical (e.g., high-energy ball milling, laser ablation) and chemical synthesis methods (e.g., chemical reduction using sodium borohydride, hydrazine, and other toxic reducing/stabilizing agents), involves hazardous chemicals and costly procedures, which cast doubts concerning the safety as well as the sustainability issue [11]. In order to eradicate this difficulty, scientists are deciding to employ green nanotechnology, at which point, they are intended to employ the utilization of plant extracts to isolate metal ions at the levels of nanoparticles in an inexpensive and eco-friendly manner. The presence of phytochemicals in the extracts presents a reducing and stabilizing effect that increases therapeutic benefits but makes them environmentally friendly [12].

Among the most promising prospects is the exploitation of underutilized medicinal plant species, the species with the ethnopharmacological potential, yet not commercial at the same time [13]. They will serve as natural nanoreactors in the production of nanoparticles, as well as possess native anti-diabetic properties [14]. A study has shown that these plant nano-formulations reduce oxidative stress, increase insulin sensitivity, and change lipid metabolism in the in vivo and in vitro tests [15]. They are more stable, specific acting, and require less dosage compared to crude extracts. It is an indication of a sustainable and strong potential of creating diabetes therapeutics through an underutilized plant and nanotechnology that come together in synergy [15, 16].

Nanoparticles of metal (silver, gold, copper, and selenium), or metal oxide (zinc oxide or ZnO), prepared by green routes were particularly shown to have great anti-diabetic properties. These nanoparticles possess antioxidant, anti-inflammatory, and glucose-regulating effects; in this regard, certain nanoparticles, such as ZnO, have demonstrated an improved insulin sensitivity and glucose uptake, but there is insufficient data to raise a conclusion concerning their safety profiles [17, 18]. Incorporation of underutilized plant extract in this technology can enhance efficacy, mitigate host toxicity, and make it affordable, a confluence of two vastly different understandings and technology [19]. This chapter acknowledges the therapeutic potential of underutilizing plant-derived metal and metal oxide nanoparticles against diabetes, pointing to their sustainability as cost-effective multifunctional agents against anti-diabetic intervention of the next-generation.

11.2 Underutilized Plants: A Phytochemical Reservoir

Neglected medicinal plants, sometimes also called orphan crops, minor crops, or underutilized species, are species of plants with known nutritional, pharmacological, or therapeutic properties but do not attract much attention in terms of usage in the form of food, or medical and health care products [20]. All of them are often limited to traditional, local, or ethnomedicinal use, and have not been part of modern pharmacopoeia or global drug discovery efforts to the same extent as such major medicinal plants as *Curcuma longa* (turmeric) or Aloe vera. Their scarcity in use is not because of therapeutic inefficiency but because of little knowledge and scientific expediency, weak market appetite, and the absence of institutional enabling interests in conservation and sustainable usage [21]. These medicinal plants can proliferate in marginal habitats and are dually well-tuned to their local agro-ecological

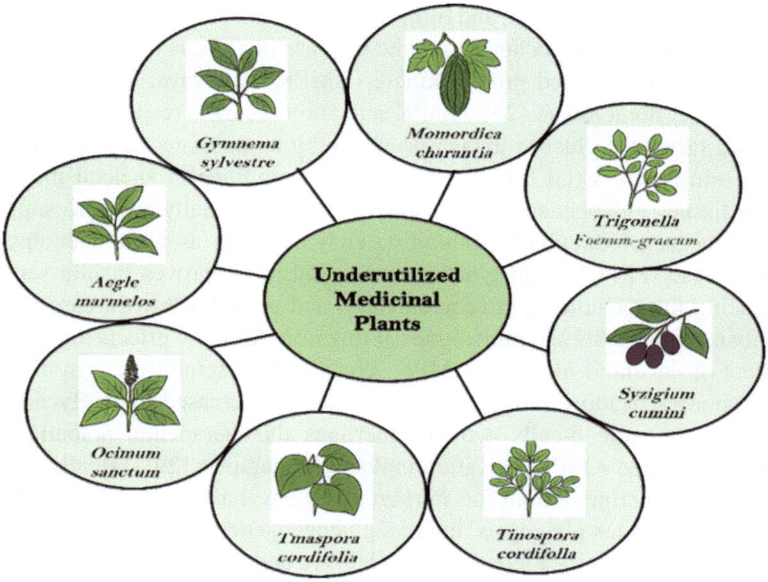

Fig. 11.1 Underutilized medicinal plants as a reservoir of bioactive phytochemicals

environment, thus presenting reservoirs of bioactive phytochemicals that have the potential to fight chronic illnesses such as diabetes, cancer, and cardiovascular diseases (see Fig. 11.1) [22].

Among the plants that are to be considered to be underutilized, the following criteria should be used: very limited geographic distribution, no large-scale cultivation, scant presence in modern pharmaceutical catalogs, and a great deal of underdevelopment of commercialization despite ethnobotanical significance [23]. Most of these plants are clinically important in the indigenous medicine systems, although they have not been subject to rigorous phytochemical and clinical scrutiny, e.g., plants such as *Momordica dioica* (spiny gourd), *Syzygium cumini* (black plum), and *Trigonella foenum-graecum* (fenugreek) have been poorly investigated despite the fact that they are clearly shown to possess anti-diabetic and antioxidant characteristics [24]. Notably, underutilized medicinal plants tend to contain secondary metabolites, such as alkaloids, flavonoids, saponins, and phenolic compounds, with therapeutic potential, which have been proven in preclinical research [25]. A more formalized way to recognize, evaluate, and assess these plants in modern phytochemical, pharmacological, and biotechnology terms could herald novel drug leads and responses to unmet medical needs. Therefore, underutilized medicinal plants can be considered as an underused, but important pool of source of future therapeutic advances [26].

Phytochemicals are naturally occurring bioactive antioxidants in plants that have been implicated in both balancing glucose levels and fighting the pathophysiological pathways of diabetes mellitus. The most investigated ones are flavonoids, which include quercetin, kaempferol, and rutin, which stimulate insulin release, protect the pancreatic beta cells, and contribute to the increase in glucose uptake due to activation of the AMP-activated protein kinase (AMPK) pathways, a key regulator of cellular energy homeostasis [27]. AMPK activation occurs in response to low energy states and promotes glucose uptake primarily by stimulating the translocation of glucose transporter 4 (GLUT4) to the plasma membrane in skeletal muscle and adipose tissue, independently of insulin [28]. Additionally, AMPK suppresses hepatic glucose production by inhibiting enzymes such as phosphoenolpyruvate carboxykinase (PEPCK) and glucose-6-phosphatase, improves insulin sensitivity by reducing intracellular lipid accumulation, and activates downstream pathways that enhance glycolysis and mitochondrial function. All these effects together cause the effect of flavonoid-mediated AMPK activation to resemble exercise-like metabolic responses, leading to increased glucose use, decreased hyperglycemia, and preserved pancreatic β-cells, which underlines the therapeutic potential of the flavonoid-mediated AMPK activation in diabetes therapies [29]. Equally, alkaloids available as berberine, within the *Berberis vulgaris*, have a potent hypoglycemic action that stimulates glycolysis, inhibits gluconeogenesis, and alters gut microbiota, evidenced by significant decreases in HbA1c and serum levels of fasting blood glucose in clinical trials [30]. Saponins, common in legumes and in underutilized plants such as *Trigonella foenum-graecum* (fenugreek), have anti-diabetes effects by slowing glucose absorption, accelerating insulin sensitivity, and reducing postprandial hyperglycemia [31]. Phenolic compounds like chlorogenic acid, readily found in *Moringa oleifera*, inhibit 6-phosphatase and glucose-alpha-glucosidase, preventing absorption of glucose in the intestine and releasing glucose into the liver [32]. Terpenoids, such as gymnemic acids in *Gymnema sylvestre*, and momordicosides in *Momordica dioica* (a little-used cucurbit) have been shown to restore β-cells and inhibit sweet taste receptors to decrease sugar cravings and enhance blood glucose levels [33]. Other potent insulin-mimetic activities and antioxidants can be found in tannins and lignans, neglected plant products that are abundant in *Syzygium cumini* (black plum), helping to prevent oxidative stress-induced pancreatic apoptosis. Polysaccharides and complex carbohydrates in certain species are prebiotic fibers that slow glucose peaks after meals and are potentially beneficial to influence gut hormones (incretins) and microbiome composition; an increasingly recognized avenue to better metabolic regulation [34].

Importantly, these phytochemicals exert their actions most likely via many mechanisms, namely, by regulating carbohydrate-digesting enzymes (alpha-amylase, alpha-glucosidase), effects upon stimulating insulin receptor signaling, defense against oxidative stress, and even regulation of inflammatory pathways such as insulin resistance [35]. The bioactive contents in many underutilized plants are exclusive combinations and therefore represent an unexploited source of antidiabetic compounds. Systematic study of these phytochemicals using third-generation metabolomics, and validation by pharmacology has the potential to provide great therapeutic arsenals of shoving diabetes (see Table 11.1) [36].

Table 11.1 Underutilized medicinal plants, their key phytochemicals, and anti-diabetic mechanisms

Plant name (underutilized)	Key phytochemicals	Anti-diabetic mechanisms	References
Helicteres isora (Indian screw tree)	Gallic acid, caffeic acid, rosmarinic acid, cucurbitacins	α-Glucosidase/α-amylase inhibition; antioxidant protection of β-cells; improved glucose tolerance	[37]
Abrus precatorius (rosary pea)	Alkaloids, flavonoids, saponins	Insulin secretagogue effects; α-amylase/α-glucosidase inhibition; antioxidant and pancreatic protective actions	[38]
Ficus deltoidei (mistletoe fig)	Flavonoids, triterpenoids, phenolics, proanthocyanidins	Enhanced insulin sensitivity and glucose uptake; antioxidant; protection against diabetic bone loss; improved tissue function	[39]
Momordica dioica (spiny gourd)	Flavonoids, cucurbitacins, saponins	Insulin sensitization; antioxidant; inhibition of intestinal enzymes; improved lipid profile	[40]
Ensete ventricosum (Ethiopian banana)	Polysaccharides, phenolics	Dietary fiber/polysaccharide slows glucose absorption; prebiotic effects; antioxidant protection	[41]
Murraya koenigii (curry leaf)	Carbazole alkaloids, flavonoids	Insulinotropic effects, antioxidant protection, inhibition of glycation	[42]
Syzygium cumini (Jamun)	Ellagic acid, tannins, flavonoids (myricetin)	α-Glucosidase inhibition; insulin mimetic effects; antioxidant and β-cell protective actions	[43]
Trigonella foenum-graecum (fenugreek)	4-hydroxyisoleucine, saponins, soluble fiber	Stimulates insulin secretion (4-hydroxyisoleucine); delays carbohydrate absorption; improves insulin sensitivity and lipid profile	[44]
Costus igneus (insulin plant)	Cysteine-rich proteins, flavonoids	Improves glycogen storage, enhances insulin sensitivity	[45]
Momordica charantia (bitter melon)	Charantin, vicine, polypeptide-p	Enhances glucose uptake, insulin mimetic effect, stimulates GLUT4 translocation	[46]
Coccinia grandis (ivy gourd)	Cucurbitacins, flavonoids, glycosides	Insulin-mimetic action; increased peripheral glucose uptake; carbohydrate-digesting enzyme inhibition	[47]

11.3 Green Synthesis of Metal and Metal Oxide Nanoparticles Using Plant Extracts

The physical and chemical methods to produce metal and metal oxide nanoparticles include laser ablation, evaporation, and condensation. The sizes of the produced metal nanoparticles range anywhere between 1 and 100 nm [48]. Alternative greener plant-derived products are used as reducing, stabilizing, and capping agents, eliminating the need to use any harsh chemicals and high-energy procedures [49]. Microorganisms and plants have evolved ways to take up and neutralize their excess metal ion, forming complex biomolecules that reduce the metal ions to matrix form. This process is less energetically demanding, cheap, and can be scaled upwards massively [50]. Nonetheless, the specific biochemical mechanism that triggers the formation of nanoparticles and parameters such as pH, temperature, length of the reaction process, and concentrations of metal ion initiation remain to be studied.

11.3.1 Plant-Mediated Synthesis of Metallic Nanoparticles

Green nanotechnology is working on the commercialization of metallic nanoparticles produced in plant-based systems. The reduction process and nanoparticle property depend on aspects of the plant extract used, metal salts, reaction conditions, and the bioactive molecules like enzymes or antioxidants present [51]. Bioactive substances interact with metal ions when metal salts are added to plant extracts, using the redox mechanism and ligand substitution mechanisms [52]. Between 2019 and 2023, over 90 plant species have been successfully used to create nanoparticles from various metals. Leaves are the most popular option due to their high concentration of secondary metabolites and their long-term harvestability, rapid regeneration, and renewable biomass supply [53]. One plant species can produce multiple types of metallic nanoparticles, such as manganese oxide and chromium oxide from *Abutilon indicum* leaf extracts [54]. In *Moringa oleifera*, distinct plant parts have been used to produce distinct nanoparticles, demonstrating the flexibility of plant-mediated synthesis in creating various nanoparticle compositions using renewable resources [55].

11.3.2 Mechanism for the Synthesis of Plant-Mediated Metal Oxide Nanoparticles

(a) Plant Selection and Extract Preparation

Green nanoparticles are obtained, but first the right species of plants with medicinal properties is extracted (anti-oxidant, anti-inflammatory, and anti-diabetic were considered). It is extracted from fresh plant materials that are obtained under healthy conditions, cleaned, and dried to preserve sensitive elements destroyed by heat [56]. A fine powder is then ground into a solvent, usually distilled water, to prevent reactive contaminants. The mixture is heated gradually between 60 and 70 °C to allow phytochemicals to seep into the solvent without degrading heat-sensitive components [57]. The extract is then filtered through muslin fabric and muslin cloth followed by Whatman filter paper to obtain a clear solution ready for use as a reducing and capping agent [50]. This process ensures the highest phytochemical content in green nanoparticles.

(b) Synthesis of Nanoparticles

Solutions of metal salts (including gold, zinc, or silver nanoparticles) are prepared in aqueous solutions of chloroauric acid ($HAuCl_4$), zinc nitrate hexahydrate ($ZnNO_3$), or silver nitrate ($AgNO_3$). The plant extract is then added slowly to the solution containing the metal salt, and a color change in the solution to the zero-valent forms of the metals is observable [58]. The natural biomolecules that occur in the extract act as natural stabilizing reagents by binding to the surface of nanoparticles, preventing aggregation and retaining the nanoparticle size. Parameters such as pH, temperature, extract-to-metal salt ratio, and time to reaction are optimized [59].

(c) Stabilization and Purification

When the synthesis is complete, the nanoparticle mixture is quickly centrifuged to pellet the nanoparticles. The nanoparticles are cleaned by shaking off the supernatant (four times), and possibly with ethanol too, to remove any salts or unbound plant metabolites. The nanoparticles are then dried in an oven, freeze or by air and left in airtight containers until needed again [60].

(d) Characterization of Nanoparticles

The synthesis of nanoparticles that has been pursued through greener synthesis techniques is confirmed through different methods of characterization to testify to the success of their synthesis, the structure, and stability of the nanoparticles. Surface plasmon resonance (SPR) is detected with UV-visible spectroscopy, whereas the crystalline nature and content of the phases are detected through X-ray diffraction (XRD) [61]. The functional groups of the phytochemicals/biomolecules that undergo reduction and capping process can be determined by Fourier-transform infrared spectroscopy (FTIR). The morphology of surface, particle size, shape, and polydispersity index are determined by both scanning electron microscopy (SEM) and transmission electron microscopy (TEM), whereas hydrodynamic size distributions are determined by DLS [62]. Surface charge and elemental composition are investigated using zeta potential analysis and energy-dispersive X-ray spectroscopy, respectively (see Fig. 11.2) [63].

11.4 Anti-diabetic Mechanisms of Underutilized Plant-Derived Nanoparticles

11.4.1 Role in Glucose Uptake and Insulin Sensitization

Nanoparticle-based materials augment peripheral glucose uptake in the tissues such as skeletal muscle and adipose tissue through regulation of glucose transporters and especially GLUT 4. The use of gold nanoparticles functionalized with propanoic acid prepared with *Cassia auriculata* enhanced glucose uptake in L6 myotubes, even in the insulin-resistant cells, thus showing an insulin-sensitizing effect [64].

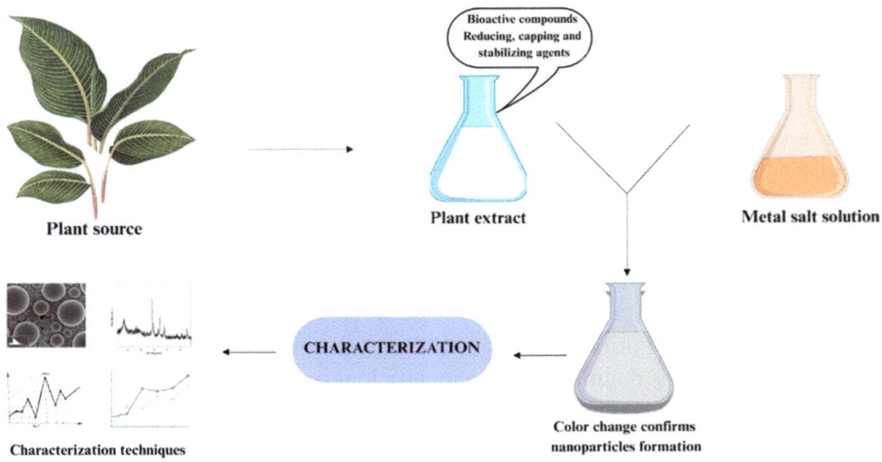

Fig. 11.2 Schematic representation of plant-derived metal and metal oxide-based nanoparticles synthesis. Plant extracts rich in phytochemicals act as natural reducing, capping, and stabilizing agents, facilitating the conversion of metal ions into biogenic nanoparticles under environmentally benign conditions

These silver nanoparticles produce a regenerating effect on the pancreatic β-cells and remodel them in dressing diabetic mice, leading to a significant reduction in fasting blood glucose [65]. This evidence suggests that nanoparticles can directly modify insulin-responsive and indirectly through pancreatic regeneration.

In support of these findings, zinc oxide nanoparticles synthesized from *Ipomoea obscura* were shown to upregulate GLUT4 expression in skeletal muscle and preserve IRS–PI3K–Akt signaling by reducing oxidative stress [17]. Selenium nanoparticles obtained by use of the *Fagonia cretica* enhanced the glycemic control in diabetic models through improvement of antioxidant enzyme activities, as well as preservation of insulin receptor signaling required to trigger the mobilization of GLUT4 [66]. Likewise, *Gongronema latifolium*, which synthesized iron oxide nanoparticles, showed both antidiabetic properties (inhibition of α-amylase and α-glucosidase) because it decreases postprandial glucose levels, and also diminishes oxidative damage that can destroy insulin sensitivity [67]. Taken together, these findings indicate that the plant-derived metal and metal-oxide nanoparticles increase glucose uptake by several mechanisms, such as direct insulin signaling stimulation, GLUT4 upregulation, oxidative stress decrease, and pancreatic β-cells regeneration.

11.4.2 Reduction of Oxidative Stress and Inflammation

Ongoing oxidative stress and long-term low-grade inflammation are major pathological characteristics of diabetes and its related complications, and they play a role in the dysfunction of the pancreatic β-cell, insulin signal impairment, and progressive tissue damage [68]. Metal and metal oxide nanoparticles, especially zinc oxide nanoparticles mediated by *Eryngium billardieri*, have demonstrated potential as promising antidiabetic nanoparticles because of their ability to reverse these pathological processes [69]. *E. billardieri* ZnO NPs showed a significant effect of restoring metabolic balance in an STZ-induced diabetic model in a rat (in adult male Wistar rats, which have the weight range of 180–220 g. Control rats with diabetes showed significant hyperglycemia, higher levels of oxidative stress enhancers, and overrated the concentration of pro-inflammatory cytokines. However, ZnO NPs also did not have this effect on rats, resulting in significant changes in fasting blood glucose levels, serum insulin levels, and redox status of the pancreas and liver.

Mechanistically, these ZnO NPs ameliorated the load of reactive oxygen species (ROS) as superoxide anions ($O_2^{-\cdot}$), hydroxyl radicals (·OH), and hydrogen peroxide (H_2O_2) through an increased activity of endogenous antioxidant enzymes as superoxide dismutase (SOD), catalase (CAT), and glutathione peroxidase (GPx) in comparison with diabetic controls that were not treated [70]. Moreover, the nanoparticles had a considerable anti-inflammatory impact, as key pro-inflammatory cytokines, especially TNF-α, IL-1β, and IL- 6, which are usually increased in diabetic control groups through the insulin resistance-related chronic inflammatory process, were significantly downregulated. Compared with diabetic controls, ZnO NP-treated rats also showed normalization of histopathological alterations in pancreatic islets and liver tissue, reflecting reduced inflammatory infiltration and oxidative damage. These findings indicate that *E. billardieri*-derived ZnO NPs disrupt the vicious cycle of ROS-induced inflammation and inflammation-induced ROS production, ultimately improving insulin sensitivity and metabolic homeostasis.

11.4.3 Enhanced Bioavailability and Targeted Delivery

The key benefit of the use of nanoparticle that is plant derived is its natural surface chemistry, and therefore increasing bioavailability and controlled release of active compounds. A large number of phytochemicals applied in the management of diabetes (e.g., quercetin, curcumin) have poor water-solubility and short life in the body. When stabilized into nanoparticles, their circulation time is prolonged and they are able to specifically accumulate in the target [71].

The natural biomolecule coating present on the plant-based nanoparticles usually serves as a capping agent and allows the plant-based nanoparticle to bind itself more selectively to the biological target to protect it [72]. Stabilization by this means not only enhances treatment results but also lowers dosages needed and decreases systemic side effects. Indicatively, plant extract formulated as nanoparticles has been reported to target pancreatic tissue more effectively, and thus stimulate β-cell production [73]. Therefore, it is important to note that the bioavailability and specific delivery of green nanoparticle has made them better choices compared to the traditional phytochemical extracts. Numerous investigations employing metal and metal-oxide nanoparticles made from genuinely underutilized plants amply confirm this advantage. For example, due to increased bioavailability and better AKT-mediated insulin signaling, zinc oxide nanoparticles (ZnO NPs) made from *Silybum marianum* seeds considerably outperformed crude extract and chemically manufactured ZnO in reducing fasting blood glucose by 28–30% [74]. Similarly, selenium nanoparticles (SeNPs) derived by using *Fagonia cretica* greatly reduced indicators of oxidative stress, increased antioxidant enzymes by 30–60%, glycemic control by 30–40%, and restored the insulin level in blood. Iron oxide (Fe_3O_4) nanoparticles prepared using *Madhuca indica* exhibited much greater amounts of enzyme interaction and cellular uptake with 73% alpha-amylase inhibition and 81% DPPH scavenging relative to the crude extract [75]. These examples indicate that phytochemical-capped nanoparticles derived by using less utilized plants are more potent antidiabetic medicines as these particles are more selective, have improved pharmacokinetics, and therapeutic efficacy compared to traditional plant extracts.

11.5 Experimental Evidences and Case Studies

The therapeutic importance of metal and metal-oxide nanoparticles composed of underutilized plants in the management of diabetes mellitus is convincingly illustrated by the growing body of experimental literature. Recent reports have indicated that phyto engineered nanoparticles have had distinct biochemical and metabolic benefits that are often superior to both that of pure plant extracts and chemically synthesized analogs, during mechanistic cell-based assays as well as carefully designed mouse models [76]. These studies can be used to synthesize bioactive nanostructures possessing enhanced stability, reactivity and biological selectivity using rare or neglected plant species through the use of nanotechnology and ethnobotanical knowledge.

These mechanistic know-hows are supported by corresponding in vivo case studies. ZnO nanoparticles in streptozotin-induced diabetic rats have been shown on multiple occasions to yield significant fasting blood glucose and glucose tolerance, serum insulin levels, and reconstitution of endogenous antioxidant enzymes, including superoxide dismutase and catalase, as well as their dual effects as an insulinotroph and an antioxidant [77]. Interestingly, ZnO-NPs prepared by *Silybum marianum* seed extract through microwave-assisted processes were found to be more potent than the crude seed extract and chemically prepared ZnO in alloxan-induced diabetic Wistar rats, which is credited to the synergistic bioactivity of the plant-derived capping molecules [17]. In addition, selenium nanoparticles that were obtained with *Fagonia cretica* extracts showed considerable enhancement in oxidative stress indicators, lipid metabolism, and inflammatory cytokines in diabetic rodent models, which further expanded the range of therapeutic uses of green-synthesized nanoparticles [78].

Comparative analysis on species such as *Madhuca indica, Tamarindus indica, Fagonia cretica*, and *Gongronema latifolium* indicates that there are common mechanistic motifs that appear despite the differences in the taxonomy of these species. These nanoparticles show recurrent antidiabetic actions such as activation of antioxidant defense systems, suppression of pro-inflammatory agents (TNF- α, IL- 6, IL-1β), inhibition of carbohydrate digestive enzymes, and reestablishment of insulin pathway signaling [75, 79, 80]. Although there are positive signs in the biology of nanoparticles produced by plant systems, a large degree of heterogeneity remains in particle size, production parameters, dose schedule, and characterization procedures [81]. Furthermore, most studies do not exhibit long-term studies, and often to a greater extent, may not be toxicologically examined to the full extent, highlighting the necessity of standardized experimental platforms to establish the full translational preparedness of plant-based nanotherapeutics.

11.6 Limitations and Future Perspectives

11.6.1 Limitations

Although underutilized metal and metal-oxide nanoparticles (NPs) derived from plants have a promising therapeutic potential in the treatment of diabetes, there are several important limitations. First, there are toxicity and safety concerns as metal and metal-oxide NPs like ZnO and Fe_3O_4 may lead to dose-dependent cytotoxicity, oxidative stress, genotoxicity, and organ accumulation, especially in the liver, kidney system, and reticulo-endothelial system [82]. The persistence, clearance, and long-term biodistribution of these NPs are not well characterized and leave doubts concerning chronic toxicity and immunity. Second, the reproducibility and standardization of green synthesis protocols is extremely diverse as there is variation in plant species, growth conditions, extraction protocols, and reaction parameters which lead to different nanoparticle sizes, shapes, surface chemistry, and capping. The fact that nanoparticles are not completely characterized (size distribution, zeta potential, surface functionalization, stability in physiological media, and ion release) further complicates the reproducibility and cross-studies comparison [83]. Third, the majority of evidence of antidiabetic activity is based on in vitro research or short-term animal models, and there are very few long-term research studies on systemic metabolism or pharmacokinetics of effects, or organ toxicity. In addition, there are practically no clinical trials in man, and safety, dosing, and efficacy in humans are guesswork. Lastly, there are still the off-target effects and the environmental issues: NPs can emit metal ions in the body, breach the unintended tissues like the brain, and either accumulate in the environment or cause ecological hazards [84].

11.6.2 Future Perspectives

To maximize the potential of plant-derived metal and metal-oxide nanoparticles in the treatment of diabetes, future studies should concentrate on rigorous safety assessment, standardization, target delivery, and translational studies. Thorough toxicology and pharmacokinetic analysis, such as biodistribution, organ distribution, immunogenicity, genotoxicity, oxidative stress, and clearance, under different dosing regimens are required [85]. Green-synthesis protocols need to be standardized in order to provide an opportunity to replicate and characterize nanoparticle size, morphology, surface chemistry, stability, and dissolution/ion release. Tissue-specific delivery (e.g., pancreatic β-cells or insulin-responsive tissues), reduction of off-target delivery, and controlled release can be improved by engineering nanoparticles (e.g., biodegradable coating, surface functionalization, and ligand-mediated targeting) [86]. The preclinical animal trials should be well designed to determine the long-term metabolism and histological effects, and then phase I/II clinical trials will be done to establish the safety, pharmacokinetics, and efficacy of the drug use in humans. Also, it is essential to avoid ecological toxicity by carrying out environmental risk assessment and designing eco-friendly nanoparticles. Incorporation of mechanistic research, computational modeling, and in vitro/in vivo correlation can further streamline the nanoparticle design process and accelerate its transfer to safe and effective antidiabetic therapeutic properties.

11.7 Conclusion

Plant-based metal and metal oxide nanoparticles (MO-NPs) are a novel and green approach to the treatment of diabetes, involving the combination of the bioactivity of underutilized medicinal plants with the intrinsic catalytic and biomedical performance of metals and metal oxides. These biogenic nanostructures provide multi-targeted intervention, which simultaneously controls hyperglycemia, oxidative stress, inflammation, and insulin resistance. MO-NPs exhibit better efficacy, improved bioavailability, and controlled release of phytochemicals as compared to crude plant extracts. Indicatively, ZnO nanoparticles prepared using *Silybum marianum* seed extract decreased fasting blood glucose by 28–30% compared to the crude extract (12–15%), whereas Fe_3O_4 nanoparticles made using *Madhuca indica* extract gave about 73–81% α-amylase inhibition and 81% DPPH scavenging, which are 45% and 52% on the crude extract.

The nano-scale size, high surface area, and cellular uptake, tissue targeting, and stability of the nanoparticles are improved by the nano-size, the high surface area, and the plant-mediated capping, permitting nanoparticles to be used in the regeneration of β-cells and the translocation of GLUT4. *Musa paradisiaca*-derived gold nanoparticles increased glucose uptake in insulin-resistant models by 25–40%. MO-NPs diminish inflammation by decreasing pro-inflammatory signaling molecules, such as TNF-α, IL-6, and IL-1β. They also reduce the levels of oxidative stress markers, including malondialdehyde (MDA), and at the same time increase activities of natural antioxidant enzymes, including superoxide dismutase (SOD), catalase, and glutathione peroxidase, hence, providing full control of the biochemical pathways that are engaged in diabetes.

By incorporating nanotechnology, phytomedicine, and AI-aid optimization, personalized and exact interventions can be designed, which reduces the side effects caused by the systemic action. Supporting green healthcare innovation: Green, scalable production with underutilized plants offers environmental friendliness. As an example, microwave-aided production of ZnO nanoparticles using the seed of *Silybum marianum* and quick aqueous production of silver nanoparticles using *Madhuca indica* leaf have been shown to synthesize stable nanostructures without using harsh chemicals, high temperatures, and toxic solvents, a proposition of a sustainable alternative to conventional nanoparticle synthesis procedures. In the future, it would be necessary to conduct long-term research in vivo, toxicological studies, and clinical translation so that MO-NPs become safe, effective, and sustainable antidiabetic drugs.

Acknowledgement The authors gratefully acknowledge the support of their institutions and colleagues in the preparation of this chapter.

Competing Interest The authors declare no competing interests.

Ethical Approval All referenced animal studies complied with institutional and international ethical guidelines; no new experiments were performed for this chapter.

References

1. Yameny AA (2024) Diabetes mellitus overview. J Biochem Res 10(3):641–645. https://doi.org/10.21608/jbaar.2024.382794
2. Agrawal P, Singh BR, Gajbe U et al (2023) Managing diabetes mellitus with millets: a new solution. J Med Life 15(9). https://doi.org/10.7759/cureus.44908
3. Cloete L (2021) Diabetes mellitus: an overview of the types, symptoms, complications and management. Nurs Sci 37(1):61–66. https://doi.org/10.7748/ns.2021.e11709
4. International Diabetes Federation (IDF) (2024) IDF. Available at: https://diabetesatlas.org/resources
5. Atlas D (2024) IDF. Available at: https://diabetesatlas.org
6. World Health Organization (WHO) (2024). Available at: www.who.int
7. Centers for Disease Control and Prevention (CDC) (2024). Available at: www.cdc.gov
8. El Mouhayyar C, Patel A, Al-Saffar N et al (2020) SGLT2 inhibitors, GLP-1 agonists, and DPP-4 inhibitors in diabetes and microvascular complications: a review. Int J Endocrinol 2020:1762164. https://doi.org/10.1155/2020/1762164
9. Shetty R, Kumar S, Rao P et al (2022) Adverse drug reactions of GLP-1 agonists: a systematic review of case reports. Diabetes Metab Syndr Clin Res Rev 16(3):102427. https://doi.org/10.1016/j.dsx.2022.102427

10. Andreadi A, Smith J, Li K et al (2024) Nanomedicine in the treatment of diabetes. Int J Mol Sci 25(13):7028. https://doi.org/10.3390/ijms25137028
11. Yadav KJ (2024) Nanotechnology in diabetes management: revolutionizing treatment and diagnostics. J Mol Liq 414:126117. https://doi.org/10.1016/j.molliq.2024.126117
12. Noah NM, Ndangili PM (2022) Green synthesis of nanomaterials from sustainable materials for biosensors and drug delivery. Sustain Chem Pharm 3:100166. https://doi.org/10.1016/j.sintl.2022.100166
13. Jauhari S, Patel R, Sharma A et al (2024) Insights on the integration of ethnopharmacology and omics in medicinal plant research. In: Ethnopharmacology and OMICS advances in medicinal plants, vol 2. Springer, pp 501–514. https://doi.org/10.1007/978-981-97-4292-9_22
14. Thirugnanasambandan T, Gopinath SC (2024) Nanoparticles from plant-based materials: a promising value-added green synthesis for antidiabetic applications. Biorefinery J 14(21):26681–26691. https://doi.org/10.1007/s13399-022-03632-5
15. Hu F, Zhang L, Wang Y et al (2022) Nanomedicine of plant origin for the treatment of metabolic disorders. Front Bioeng Biotechnol 9:811917. https://doi.org/10.3389/fbioe.2021.811917
16. Stoleru OA, Popescu MN, Ionescu C et al (2024) Multiple nanotechnological approaches using natural compounds for diabetes management. Diabetes Metab Syndr Clin Res Rev 23(1):267–287. https://doi.org/10.1007/s40200-023-01376-1
17. Yousaf I (2024) The current and future perspectives of zinc oxide nanoparticles in the treatment of diabetes mellitus. J Adv Pharm Appl 12. https://doi.org/10.48550/arXiv.2409.04486
18. Bhardwaj M, Singh P, Choudhury R et al (2020) A review on ameliorative green nanotechnological approaches in diabetes management. Biomed Pharmacother 127:110198. https://doi.org/10.1016/j.biopha.2020.110198
19. Taha NA, Ali S, Hassan H et al (2025) A mini-review of nanoparticle therapeutics targeting oxidative stress and inflammation in diabetes. J Diabetes Res 86(1):29. https://doi.org/10.1186/s41936-025-00449-2
20. Padulosi S, Thompson J, Rudebjer P et al (2013) Fighting poverty, hunger and malnutrition with neglected and underutilized species: needs, challenges and the way forward. https://doi.org/10.13140/RG.2.1.3494.3842
21. Mayes S, Jones D, Kumar P et al (2012) The potential for underutilized crops to improve security of food production. J Exp Bot 63(3):1075–1079. https://doi.org/10.1093/jxb/err396
22. Macha MA, Singh R, Gupta V et al (2025) Neglected and underutilized crops: unveiling potential for cancer drug discovery. https://doi.org/10.1201/9781003477792
23. Borelli T, Kumar A, Singh P et al (2023) Inclusion criteria for underutilized food plants in nutrition-sensitive programming. In: Neglected Plant Foods of South Asia. Springer, pp 73–100. https://doi.org/10.1007/978-3-031-37077-9_4
24. Coopoosamy R, Naidoo P, Govender T et al (2025) The role of phytomedicine: bridging the gap between the past, present, and future. https://doi.org/10.48741/mut.24182394
25. Shahrajabian MH, Sun W (2024) The power of the underutilized and neglected medicinal plants and herbs of the Middle East. Rev Recent Clin Trials 19(3):159–175. https://doi.org/10.2174/0115748871276544240212105612
26. Bhatti MZ, Ismail H, Kayani WK et al (2022) Plant secondary metabolites: therapeutic potential and pharmacological properties. In: Secondary metabolites — trends and reviews. IntechOpen. https://doi.org/10.5772/intechopen.103698
27. Akpoveso O-OP, Ubah EE, Obasanmi G et al (2023) Antioxidant phytochemicals as potential therapy for diabetic complications. Antioxidants (Basel) 12(1):123. https://doi.org/10.3390/antiox12010123
28. Steinberg GR, Hardie DG (2023) New insights into activation and function of AMPK. Nat Rev Mol Cell Biol 24(4):255–272. https://doi.org/10.1038/s41580-022-00547-x
29. Han S, Lee J, Kim H et al (2023) Dietary flavonoids prevent diabetes through epigenetic regulation: advances and challenges. Crit Rev Food Sci Nutr 63(33):11925–11941. https://doi.org/10.1080/10408398.2022.2097637

30. Baska A, Singh R, Sharma P et al (2021) Berberine in the treatment of diabetes mellitus: a review. Curr Drug Metab 21(8):1379–1386. https://doi.org/10.2174/1568026620666201022144405
31. Trasca DM, Ionescu C, Popescu MN et al (2025) The antidiabetic activity of wild-growing and cultivated medicinal plants used in Romania: a review. Pharmaceuticals 18(7):1035. https://doi.org/10.3390/ph18071035
32. Sun W, Shahrajabian MH (2023) Therapeutic potential of phenolic compounds in medicinal plants. Molecules 28(4):1845. https://doi.org/10.3390/molecules28041845
33. Bhatt SM, Sharma N, Verma P et al (2023) The potential of various herbal plants for therapeutic diabetes treatment. Ann Natl Acad Med Sci:1–11. https://doi.org/10.25259/ANAMS_70_2024
34. Li F, Wang J, Zhang Y et al (2025) Lowering glycemic levels via gastrointestinal tract factors. Crit Rev Food Sci Nutr 65(3):575–611. https://doi.org/10.1080/10408398.2023.2278169
35. Aryal D, Shrestha S, Rai K et al (2024) Dietary phenolic compounds as promising therapeutic agents for diabetes. Food Sci Nutr 12(5):3025–3045. https://doi.org/10.1002/fsn3.3983
36. Surendran K, Nair R, Raghavan V et al (2023) Genomics and metabolomics in medicinal plant pathways. In: Phytochemical genomics. Springer, pp 343–360. https://doi.org/10.1007/978-981-19-5779-6_13
37. Chavre BJ (2024) Helicteres isora L.: a traditional medicinal plant. Int J Pharm Environ 10(03):179–183. https://doi.org/10.18811/ijpen.v10i03.23
38. Dafar S, Khan M, Patel R et al (2023) A review on phytochemical and pharmacological study of Abrus precatorius. Int J NnrPh:161–173. https://doi.org/10.61554/ijnrph.v1i2.2023.31
39. Ashraf K, Ahmed S, Khan T et al (2021) Overview of phytochemical and biological activities of Ficus deltoidea. J Pharm Bioallied Sci 13(1):11–25. https://doi.org/10.4103/jpbs.JPBS_232_19
40. Bhoye SJ (2021) Spine Gourd (Momordica dioica). Agric Food Newsl 3:409–411
41. Banadka A, Kumar S, Singh R et al (2024) Bioactive compounds and biological activities of Ensete species. In: Bioactive compounds in plant storage organs. Springer, pp 77–103. https://doi.org/10.1007/978-3-031-44746-4_4
42. Tabashiri A, Mohammadi M, Rahimi R et al (2022) A decade of anti-diabetic potential of Murraya koenigii. Front Nutr 30(3). https://doi.org/10.54931/2053-4787.30-3-2
43. Pujiastuti E, Wibowo A, Nugroho H et al (2023) Contribution of phytochemicals in *Syzygium cumini* as antidiabetics: a systematic review. Int J Pharm 34(4). https://doi.org/10.22146/ijp.6812
44. Sarker DK, Hossain M, Rahman T et al (2024) Antidiabetic potential of fenugreek (Trigonella foenum-graecum). Food Sci Nutr 12(10):7108–7136. https://doi.org/10.1002/fsn3.4440
45. Mathew F, Varghese B (2019) Medicinal exploration of *Costus igneus*: the insulin plant. 54(2):51–57
46. Banerjee J, Chanda R, Samadder A (2019) Anti-diabetic activity of *Momordica charantia*: a review. ASPS 3:24–30
47. Jamwal A, Kumar S (2019) Antidiabetic activity of isolated compound from Coccinia indica. Int J Pharm Epidemiol Res 53(1):151–159. https://doi.org/10.5530/ijper.53.1.20
48. Chen Y, Li X, Zhang H et al (2020) Phase engineering of nanomaterials. Chem Rxiv 4(5):243–256. https://doi.org/10.1038/s41570-020-0173-4
49. Rónavári A, Kovács D, Balogh L et al (2021) Green silver and gold nanoparticles. Molecules 26(4):844. https://doi.org/10.3390/molecules26040844
50. Hussain A, Khan M, Shah S et al (2019) Biogenesis of ZnO nanoparticles using Pandanus odorifer. RSC Adv 9:15357–15369. https://doi.org/10.1039/C9RA01659G
51. Aboyewa JA, Ndlovu S, Moyo T et al (2021) Green synthesis of metallic nanoparticles using medicinal plants of southern Africa. Plants (Basel) 10(9):1929. https://doi.org/10.3390/plants10091929
52. Shafey AME (2020) Green synthesis of metal and metal oxide nanoparticles from plant leaf extracts. Green Process Synth 9(1):304–339. https://doi.org/10.1515/gps-2020-0031

53. Aslam M, Khan R, Ahmed S et al (2021) Green synthesis of TiO_2 nanoparticles using plant biomolecules. J Ind Eng Chem 98:1–16. https://doi.org/10.1016/j.jiec.2021.04.010
54. Bao Y (2021) Plant-extract-mediated synthesis of metal nanoparticles. Adv Sustain Syst 2021:6562687. https://doi.org/10.1155/2021/6562687
55. Jayachandran A, Ramesh S, Kumar P et al (2021) Green synthesis and characterization of ZnO nanoparticles using Cayratia pedata. Biochem Biophys Rep 26:100995. https://doi.org/10.1016/j.bbrep.2021.100995
56. Farshchi HK, Mohammadi M, Rahimi R et al (2018) Green synthesis of iron nanoparticles by rosemary extract. Biotechnol Prog 16:54–62. https://doi.org/10.1016/j.bcab.2018.07.017
57. Dagdag O, Patel H, Singh R et al (2024) Phytochemicals/plant extracts: preparation and properties. In: Phytochemistry in corrosion science. CRC Press, pp 20–36. https://doi.org/10.1201/9781003394631
58. Subhasree S, Ramesh S, Kumar A et al (2020) Anticorrosion behavior of ZnO nanoparticles. J Nano-Nano Sci 20(7):4061–4068. https://doi.org/10.1166/jnn.2020.17526
59. Demirezen DA, Kaya H, Yildiz O et al (2019) Green synthesis of iron oxide nanoparticles using Ficus carica. J Biosci 127(2):241–245. https://doi.org/10.1016/j.jbiosc.2018.07.024
60. Alsammarraie FK, Ahmed S, Khan R et al (2018) Green synthesis of silver nanoparticles using turmeric. Colloids Surf B Biointerfaces 171:398–405. https://doi.org/10.1016/j.colsurfb.2018.07.059
61. Can MJ (2020) Green gold nanoparticles from plant materials. Rev Inorg Chem Eng 36(7):859–877. https://doi.org/10.1515/revce-2018-0051
62. Devi HS, Ramesh S, Kumar P et al (2019) Green synthesis of iron oxide nanoparticles using Platanus orientalis. Green Process Synth 8(1):38–45. https://doi.org/10.1515/gps-2017-0145
63. Wadhwa P, Sharma R, Verma A et al (2022) A review of nanoparticle characterization techniques. Adv Nanotechnol 7(3):202–214. https://doi.org/10.2174/2405461507666220405113715
64. Govindaraju K, Suganya KU, Kumar S et al (2020) Anti-diabetic assessment of guavanoic acid functionalized gold nanoparticles. Medchem Commun 11(7):814–822. https://doi.org/10.1039/D0MD00125B
65. Wahnou H, Tchoumbougnang F, Moukette BM et al (2025) Curcumin-based nanoparticles: advancements in tumor therapy. Pharmaceutics 17(1):114. https://doi.org/10.3390/pharmaceutics17010114
66. Satpathy S, Panigrahi LL, Arakha M et al (2024) Selenium nanoparticles in diabetic complications: a comprehensive study. Curr Top Med Chem 24(15):1327–1342
67. Ekozin A, Oladele O, Abiola F et al (2024) Green synthesized iron nanoparticles using Gongronema latifolium. 9(1):137–149
68. Dludla PV, Nkambule BB, Silvestri S et al (2023) Pancreatic β-cell dysfunction in type 2 diabetes. World J Diabetes 14(3):130. https://doi.org/10.4239/wjd.v14.i3.130
69. Kazempour P, Salehi M, Khorsandi L et al (2021) Green synthesis of ZnO nanoparticles using *Eryngium billardieri* leaf extract: characterization and its anti-diabetic properties. Mol Biol 7:18
70. Lin L-S, Zhang H, Wang Y et al (2019) Zinc peroxide nanoparticles enhance oxidative stress therapy. Theranostics 9(24):7200. https://doi.org/10.7150/thno.39831
71. Sarvarian P, Fathi M, Hosseini S et al (2022) Emerging plant-derived nanoparticles for nano-drug delivery. Drug Deliv Transl Res 51(4):1039–1059. https://doi.org/10.1080/08820139.2021.1891094
72. Patel S, Kumar R, Sharma P et al (2022) Plant-based bioavailability enhancers. Curr Drug Deliv 28(8):642–654. https://doi.org/10.2174/1381612828666220112141355
73. Wang M, Li J, Chen Y et al (2022) Targeted polymeric nanoparticles based on mangiferin. ACS Appl Mater Interfaces 14(9):11092–11103. https://doi.org/10.1021/acsami.1c22964
74. Arvanag FM, Hosseini H, Jamalifar H et al (2019) Antidiabetic and antibacterial activities of ZnO nanoparticles from Silybum marianum. Mater Sci Eng C 97:397–405. https://doi.org/10.1016/j.msec.2018.12.058

75. Shabbir MA, Khan R, Hussain S et al (2023) Iron oxide nanoparticles from Madhuca indica: antidiabetic properties. ACS Omega 8(37):33358–33366. https://doi.org/10.1021/acsomega.3c02744
76. Radulescu D-M, Popescu C, Ionescu R et al (2023) Green synthesis of metal and metal oxide nanoparticles: biomedical applications. Int J Mol Sci 24(20):15397. https://doi.org/10.3390/ijms242015397
77. Afify M, El-Sayed W, Ali S et al (2019) Effect of ZnO nanoparticles on diabetes in rats. Eur J Chem 62(10):1771–1783. https://doi.org/10.21608/ejchem.2019.11350.1735
78. Khan HA, Khan R, Siddiqui S et al (2023) Antidiabetic evaluation of selenium nanoparticles from Fagonia cretica. Biomed Pharmacother 164:114872. https://doi.org/10.1016/j.biopha.2023.114872
79. Khan HA, Khan R, Ali S et al (2023) Green synthesis of silver nanoparticles from Fagonia cretica. Front Pharmacol 14:1194809. https://doi.org/10.3389/fphar.2023.1194809
80. Kwatra A, Singh P, Gupta R et al (2025) Green ZnO nanoparticle synthesis using Tamarindus indica. Spectrochim Acta A Mol Biomol Spectrosc 6:100083. https://doi.org/10.1016/j.scca.2025.100083
81. Lithi IJ, Monga R, Patel A et al (2025) Review on green synthesis of metal and metal oxide nanoparticles. Nanoscale Adv. https://doi.org/10.1039/d5na00037h
82. Sengul AB, Asmatulu E (2020) Toxicity of metal and metal oxide nanoparticles. Environ Sci Pollut Res Int 18(5):1659–1683. https://doi.org/10.1007/s10311-020-01033-6
83. Djurišić AB, Leung YH, Ng A et al (2015) Toxicity of metal oxide nanoparticles: mechanisms and characterization. Small 11(1):26–44. https://doi.org/10.1002/smll.201303947
84. Lead JR, Wilkinson K, Jager T et al (2018) Nanomaterials in the environment. Environ Toxicol Chem 37(8):2029–2063. https://doi.org/10.1002/etc.4147
85. Tripathi D, Sharma R, Verma P et al (2024) Advances in nanomaterials for precision drug delivery. Biomed Res India 15:30573. https://doi.org/10.34172/bi.30573
86. Tiwari R, Singh S, Gupta A et al (2025) Nanogenetics in diabetes: nanoscale platforms and gene modulation for metabolic reprogramming. J Biomater Sci Polym Ed:1–52. https://doi.org/10.1080/09205063.2025.2591372

Chapter 12
Role of Underutilised Plant-Derived Metal and Metal Oxide-Based Nanoparticles Against Obesity

Vandna Bhardwaj, Manikanika, Bhavna Patial, and Nidhi Bhardwaj

Abstract Obesity, a growing global health concern, is a significant risk factor for several chronic diseases, including non-alcoholic fatty liver disease (NAFLD), cardiovascular disorders, cancer, diabetes and immune dysfunction. Impaired metabolic regulation is often fuelled by high calorie diet and a sedentary lifestyle, which leads to excessive fat accumulation and oxidative stress. The imbalance between pro-antioxidants and antioxidants, along with chronic live tissue inflammation, aggravates the progression of NAFLD and other obesity related complications. Recent advances in nanotechnology have introduced a promising therapeutic approach using metal and metal oxide-based nanoparticles (NPs), especially those derived from underutilised medicinal plants. Strong antioxidant, anti-inflammatory and lipid-lowering qualities make these plant-mediated NPs ecofriendly and biocompatible substitute for traditional obesity treatments. The increasing relevance of plant-derived metal and metal oxide nanoparticles for the treatment of obesity and other related metabolic dysfunctions is discussed, offering a fresh, sustainable and affordable approach to this prevalent health issue.

Keywords Antioxidant · Metal-oxide · Nanoparticles · Obesity · Metabolic dysfunction

---

V. Bhardwaj · B. Patial
School of Biological and Environmental Sciences, Shoolini University of Biotechnology and Management Sciences, Solan, India

Manikanika
Department of Chemistry, University Institute of Sciences (UIS), Chandigarh University, Gharuan, Mohali, Punjab, India

N. Bhardwaj (✉)
Center of Advanced Innovation Technologies, VŠB-Technical University of Ostrava, Ostrava-Poruba, Czech Republic

Department of Chemistry, Faculty of Science, University of Hradec Kralove, Hradec Kralove, Czech Republic

© The Author(s), under exclusive license to Springer Nature Switzerland AG 2026

E. Nepovimova et al. (eds.), *Underutilized Plants Based-Nanoparticles: Herbal Remedy to Treat Human Diseases*, https://doi.org/10.1007/978-3-032-20384-7_12

Abbreviations

AgNPs	Silver nanoparticles
AuNPs	Gold nanoparticles
BMI	Body mass index
C/EBP	CCAAT/Enhancer binding protein
CuNPs	Copper nanoparticles
CuONPs	Copper oxide nanoparticles
DLS	Dynamic light scattering
DNA	Deoxyribose nucleic acid
Fe_3O_4-NPs	Iron oxide nanoparticles
FTIR	Fourier transform infrared spectroscopy
HFD	High fat diet
kg	Kilogram
LEP	Leptin gene
MC4R	Melanocortin 4 receptor gene
mg	Milligram
MgO-NPs	Magnesium oxide nanoparticles
mL	Millilitre
NAFLD	Non-alcoholic fatty liver disease
nm	Nanometre
NPs	Nanoparticles
NPs	Nanoparticles
NUS	Neglected and underutilised species
PCOS	Polycystic ovarian syndrome
PCSK1	Proprotein convertase subtilisin/kexin type 1
POMC	Pro-opiomelanocortin gene
PPARγ	Peroxisome proliferator-activated receptor gamma
SEM	Scanning electron microscopy
TEM	Transmission electron microscopy
TiO_2	Titanium dioxide
UV	Ultraviolet
WHO	World Health Organisation
XRD	X-ray diffraction
ZnO-NPs	Zinc oxide nanoparticles

12.1 Introduction

Obesity has emerged as one of the most obstinate global health concerns of the twenty-first century. Over the past three decades, its occurrence has drastically increased, significantly exceeding the advancements in therapeutic methods and research. Earlier considered merely as a result of poor lifestyle choices, obesity is now recognised to be a complex disorder shaped by biological, psychological and social determinants [1]. Type 2 diabetes, hypertension, dyslipidaemia, sleep apnoea, osteoarthritis and various types of cancer are all closed linked to this chronic condition which results in disability, reduced productivity and an increased healthcare burden worldwide. The intricacy and enduring nature of the epidemic highlight the pressing need for efficient, secure and long-lasting approaches to the fight against obesity. The most often used tool for diagnosis of obesity is the body mass index (BMI), calculated by dividing body weight (in kilograms) by the square of height (in meters). A BMI of 30 or above is generally categorised as obesity. Despite its utility in population-level screening, BMI does not accurately reflect fat distribution or body composition. As a result, the World Health Organisation (WHO) recommends its use primarily as a screening tool rather than a definitive diagnostic measure [2]. The introduction of nanotechnology in the research of obesity and related problems has promised the development of efficient and targeted methods for treatment. The present work highlights the current advancements of nanoparticles for obesity management, green synthesis of nanoparticles, the mechanism of action, therapeutic potential and further challenges to overcome for implementation.

12.2 Aetiology and Pathophysiology of Obesity

Obesity is a complex condition that is influenced by interactions of genetic, environmental and psychosocial factors (Fig. 12.1). The primary driver remains long-term energy imbalance, where caloric intake consistently exceeds expenditure. The surplus energy is stored in adipose tissue, leading to fat cell hypertrophy and metabolic alterations that perpetuate obesity-related pathology [3, 4].

12.2.1 Genetic Contributions

Genetic modifications play an important role in obesity. Mutations, DNA polymorphisms and altered gene expression can influence appetite regulation, fat storage and metabolic efficiency. Obesity is typically classified into three main genetic types: monogenic, polygenic and syndromic. Monogenic obesity is rare but more severe and usually presents in childhood due to mutations in single genes like MC4R (melanocortin 4 receptor gene), LEP (leptin gene), POMC (pro-opiomelanocortin gene), PCSK1 (Proprotein convertase subtilisin/kexin type 1) [5–7]. Polygenic obesity is most prevalent and arises from the cumulative effect of multiple variants. Syndromic obesity is associated with developmental abnormalities and organ dysfunction [8, 9].

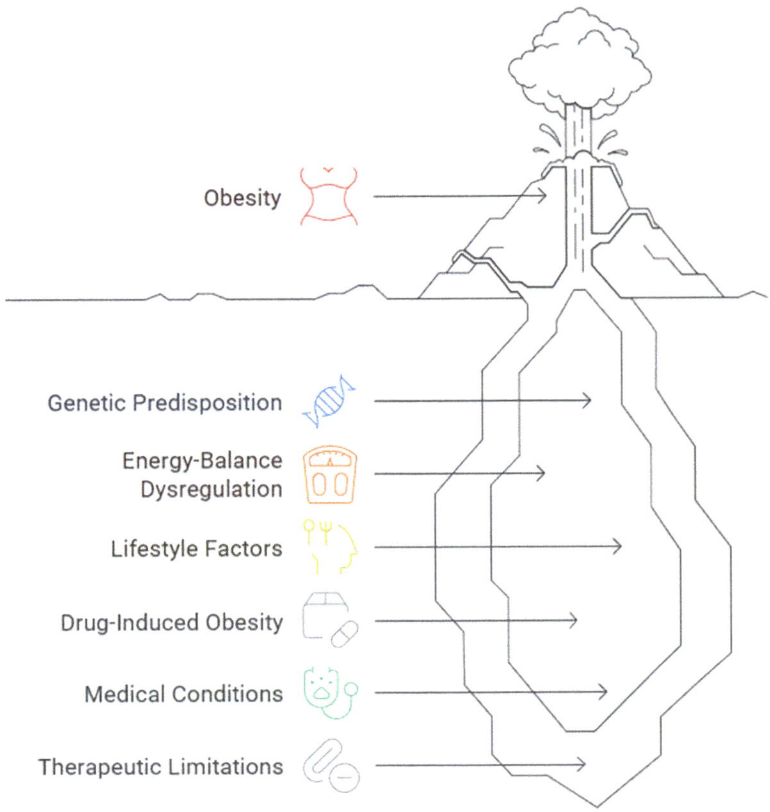

Fig. 12.1 The deep rooting causes and contributing factors for obesity

12.2.2 Energy-Balance Dysregulation

Energy balance is regulated by central and peripheral mechanisms involving the hypothalamus, adipose tissue, gut, pancreas and microbiome. The hypothalamus plays a major role in regulating appetite and energy use. In addition to the hypothalamus, other regions of the brain also influence eating through hedonic food responses, sensory input and cognitive regulation.

When people try to lose weight, the body often fights back by lowering energy use and increasing hunger signals. This makes it hard to keep weight off and explains why many people regain weight. Because of this, obesity is considered a long-term, recurring disease that needs lifelong care [3].

12.2.3 Lifestyle and Behavioural Factors

- Physical Inactivity: Sedentary behaviours, facilitated by urbanisation and passive transportation, are major contributors. The COVID-19 lockdowns further exacerbated inactivity on a global scale [1].
- Sleep Deprivation: Poor sleep affects neuroendocrine function, lowering leptin and insulin sensitivity while elevating cortisol`and ghrelin levels, thereby increasing appetite [10]. Sleep and exercise are interlinked, with disturbances in one often affecting the other [11].

12.2.4 Drug-Induced Obesity

Several therapeutic agents predispose individuals to weight gain. Antidepressants, antipsychotics and certain antidiabetic medications can alter appetite, food cravings and glucose metabolism, leading to increased adiposity [12–14].

12.2.5 Medical Conditions

Endocrine and metabolic disorders contribute significantly to obesity. Hypothyroidism, Cushing's syndrome and polycystic ovary syndrome (PCOS) are well-known causes. Rare conditions such as insulinomas also promote weight gain through excess insulin secretion. Additionally, metabolic syndrome—a cluster of hypertension, insulin resistance, abdominal obesity and dyslipidaemia—further compounds obesity risk [1].

12.2.6 Current Therapeutic Approaches and Limitations

Conventional pharmacotherapies for obesity, including appetite suppressants and metabolic modulators, often suffer from limited efficacy, poor bioavailability and off-target effects. Adverse outcomes have led to the withdrawal of several anti-obesity drugs like Rimonabant (due to psychiatric side effects) Fenfluramine and Dexfenfluramine (due to primary pulmonary hypertension and heart valve problems), etc., from the market [15]. These limitations have shifted focus towards safer, naturally derived alternatives [16, 17].

12.3 Role of Neglected and Underutilised Plant Species (NUS)

The exploration of neglected and underutilised species (NUS) has gained momentum in recent years as a sustainable approach for food security and novel therapeutics. Defined as non-commodity crops that were once more widely used but are now marginalised due to agronomic, economic and cultural reasons, NUS represent a valuable reservoir of bioactive compounds [18].

Several NUS have demonstrated anti-obesity potential:

- *Solanum aethiopicum*: Ethanolic extracts obtained by exhaustive maceration reduced body weight gain, improved lipid profiles and lowered inflammatory markers in high-fat diet (HFD) mice, suggesting direct anti-obesity effects [19].
- *Crithmum maritimum* (Sea Fennel): Traditionally used for weight control, it contains polyphenols and essential oils that may serve as functional food ingredients [20].
- *Moringa peregrina*: Leaf and bark extracts reduced fat accumulation, oxidative stress and inflammation while improving glucose tolerance and liver function in HFD-fed rats [21].
- *Bidens pilosa* (Blackjack): A wild leaf vegetable often neglected in many regions. Studies have shown that the extract of *B. pilosa* helps in reducing body weight, visceral fat and adipogenesis by downregulating transcription factors such as C/EBP and PPARγ. It also shows the antidiabetic and alpha-amylase inhibitory effects in animal models and in vitro assays [22].
- *Moringa oleifera*: It is widely known nutritionally, but underutilised for therapeutic purposes in many regions. Leaf and seed extracts prevent weight gain in high-fat diet animal models, improve lipid metabolism and positively modulate gut microbiota [23].
- *Amaranthus* spp.: It is often consumed in some regions as leafy vegetables or pseudocereals, but remains underutilised globally. Studies have shown that *Amaranthus* supplementation improves hepatic lipid profiles, reduces triglycerides and cholesterol and enhances glycaemic control [24].
- *Salicornia arabica: Salicornia* extract obtained from its aerial parts significantly reduced body weight and improved lipid profiles, highlighting its potential as a functional food [25].

These findings highlight the potential of underutilised plants in addressing obesity through mechanisms such as appetite regulation, adipogenesis inhibition, thermogenesis stimulation and enhancement of insulin sensitivity.

Over the past few years, nanotechnology has been gaining significant attention in the field of biomedical research by introducing new diagnostic techniques to treat complex problems, including obesity. Nanoparticles have shown their remarkable physical and chemical properties, which include their smaller size, greater surface area, controllable surface chemistry and improved bioavailability. They have shown improvement in drug solubility, drug delivery, facilitate cell transport and extended release of bioactive agents [26]. In spite of this, nanoparticles have been explored for distinct applications in the management of obesity, introduction and delivery of drugs required for anti-obesity, blockage of adipogenesis and to monitor the distribution of adipose tissues [27]. Nanotechnology has shown advancements in obesity management with the help of nanoparticles, specifically metal oxide nanoparticles (Table 12.1). Nanoparticles have distinct properties which is responsible for therapeutic interventions. Nanoparticles, which are plant-based and synthesised from the underutilised plant extract have become green approach nanoparticles synthesis [28]. The phytochemical components present in plant extract act as reducing agents, which stabilise and make the bioactivity of nanoparticles functional [29]. This synthesis is not only environmentally friendly but also combines the medicinal properties of plants and the physicochemical advantages of metal oxides.

12.4 Plant-Based Metal Nanoparticles and Metal Oxide Nanoparticles

There are numerous underutilised plants which are rich in flavonoids, alkaloids, polyphenols and terpenoids, which are helpful in obesity and diabetes management. The green synthesis approach has become a promising method than other physical and chemical methods due to its environmentally sustainable approach. This method not only decreases the toxic chemicals from the environment, but also provides nanoparticles with surface modifications and biological compatibility. In green synthesis, it involves the mixing of metal precursors with their plant extract derived from their leaves, roots, seeds, etc. The components which are present act as reducing agents and convert metal ions to their oxides, and also, they act as capping agents to prevent nanoparticles from aggregation. Various parameters like temperature, extract concentration and pH affect their morphology and other surface properties. Nanoparticles synthesised typically range from 10 nm to 100 nm and become stable (Table 12.1) [30].

12.4.1 Types of Plant-Derived Metal Oxide Nanoparticles

(a) Zinc Oxide Nanoparticles (ZnO-NPs)

Zinc oxide nanoparticles are extensively explored metal oxides derived from plant extracts due to their broad spectrum of bioactivity. Various plant sources, such as *Aloe vera* [31], *Deverra tortuosa* [32], *Atalantia monophyla* [33], *Hibiscus rosa-sinensis* [34], are used to synthesise zinc oxide nanoparticles as they are the natural source of reducing agents in the form of terpenoids, flavonoids and phenolics, which are helpful to stabilise nanoparticles. These nanoparticles have strong antioxidant and antimicrobial actions, which can be used in biomedical applications. These

Table 12.1 Plant-extract–mediated nanoparticles (NPs) investigated for anti-obesity effects, including source, dose/model, key phytochemicals, observed outcomes and nanoparticle characteristics

Sr. No.	NPs	Plant extract used	Dose/catalyst used to study obesity (unit & model)	Phytochemicals (linked to anti-obesity)	Model & main observed effect	NP notes (size/ synthesis)	Reference
1.	AgNPs	*Persea americana*	In vitro 3T3-L1 adipogenesis: 0.5–2.5 mg/mL (dose-dependent inhibition). In vivo HFD rat work reported (see methods for mg/kg)	Tannins, flavonoids, steroids, saponins, phenolics (anti-lipogenic)	In vitro: strong inhibition of adipocyte differentiation In vivo: improved lipid profile and adipose histology reported	AgNPs spherical; characterised by UV/TEM/SEM. Often tested with drug co-load (e.g. orlistat) in some studies	[48]
2.	AuNPs	*Panax ginseng*/*Gynostemma pentaphyllum* (leaf extracts)	In vitro 3T3-L1: effective at ~ 50–100 µg/mL (anti-adipogenic assays); some. In vivo AuNP studies used daily IP doses (µg–mg/kg range)	Ginsenosides/ gypenosides, polysaccharides, saponins— anti--inflammatory & metabolic regulators	In vitro: ↓lipid accumulation, downregulation of PPARγ/CEBPα and adipogenic genes In vivo (rodents): AuNPs reduced fat mass, improved glucose and lipid parameters	AuNPs ~ 10–30 nm; green synthesis with leaf extracts; characterised by UV/TEM/XRD	[49]

3.	ZnO NPs	*Dendropanax morbifera*	In vitro 3T3-L1: effective concentrations reported around 10 μg/mL for reduced lipid accumulation; some formulations tested in nanoemulsions for improved delivery	Polyphenols, flavonoids, antioxidant compounds (phenolics implicated in anti-adipogenic effect)	In vitro: ↓lipid accumulation, enhanced lipolysis, downregulation of adipogenic markers; used in co-delivery (e.g. tryptophan) nanoformulations to boost efficacy	ZnO NPs characterised (DLS/TEM); nanoemulsion hydrodynamic sizes ~ 140–150 nm for formulations	[50]
4.	(Cu/CuO NPs)	*Bauhinia variegata*	In vitro lipase/amylase inhibition assays: concentrations from 20–100 μg/mL (lipase inhibition % reported); some works report further. In vitro anti-obesity metrics	Flavonoids, tannins, saponins, phenolics (enzyme inhibition & antioxidant activity)	In vitro: significant pancreatic lipase inhibition and amylase inhibition (anti-obesity/anti-digestive enzyme activity); suggested for further. In vivo testing.	Green CuNPs/CuO NPs sizes variable (8–100 nm reported depending on plant and method); characterised by UV/DLS/SEM/XRD	[51]

(continued)

Table 12.1 (continued)

Sr. No.	NPs	Plant extract used	Dose/catalyst used to study obesity (unit & model)	Phytochemicals (linked to anti-obesity)	Model & main observed effect	NP notes (size/synthesis)	Reference
5.	TiO_2 NPs	*Bauhinia rufescens*	In vitro (maturing adipocytes): tested in adipogenesis assays (concentrations as per Methods; Lipid accumulation & adipokine release measured)	Bark extract phytochemicals (flavonoids, tannins)	In vitro (3T3-L1): TiO_2-NPs loaded with plant extract inhibited lipid accumulation and reduced pro-inflammatory adipokines—candidate for anti-obesity research	TiO_2 NPs loaded with plant extract; characterised (size, FTIR, XRD); used to improve delivery of phytocompounds	[52]
6.	AgNPs	*Ziziphus spina-christi*	In vivo/ex vivo: studies reported anti-obesity--relevant molecular effects. In vitro adipocyte assays (3T3-L1) with various µg/mL doses (check Methods for exact concentrations)	Alkaloids, saponins, flavonoids, phenolics—reported anti-inflammatory & metabolic activity	In vitro & ex vivo: reduced lipid accumulation, modulated inflammatory markers related to metabolic syndrome	AgNPs biosynthesised with Ziziphus extract; characterised by UV/TEM/FTIR; sizes reported in the paper	[53]

nanoparticles have emerged as a nanotherapeutic to manage obesity as they can control glucose homeostasis by the release of insulin, translocation of glucose transporter and pancreatic beta cell activity [35].

(b) Magnesium Oxide Nanoparticles (MgO-NPs)

MgO-NPs are also playing an important role in biomedical applications. They are also synthesised by using the green method by using distinct plant sources such as *Tarenna asiatica* [36], *Trigonella foenum-graecum* [37] and *Abrus precatorius* L. [38]. These sources are rich in polyphenols and organic acids, which regulate metabolic activities because magnesium acts as a cofactor in insulin signalling, metabolising carbohydrate and lipid homeostasis. The nanoparticles derived from plants have an ability to enhance glucose uptake in peripheral tissues and also regulate serum lipid profiles which reduces the hyperglycaemia and obesity-induced dyslipidaemia [39]. They contain a natural phytochemical coating, which, in addition to improving bioavailability and reducing cytotoxicity, makes them appropriate in long-term therapeutic regimens.

(c) Iron Oxide Nanoparticles (Fe_3O_4-NPs)

Iron oxide nanoparticles can also be synthesised by using various plant sources, which extract distinctive magnetic and antioxidant properties that play a role in drug delivery and metabolic regulation. Plants like *Lagenaria siceraria* [40] and *Avicennia marina* [41] have phytochemicals such as ellagic acid, tannins and catechins, which reduce Fe^{3+} ions and make a stabilised organic layer around nanoparticles, which further helps Fe_3O_4-NPs to target drug delivery to adipose tissues and liver, particularly in obesity related disorders. Also, they are capable of minimising cytokine driven inflammation and oxidative stress to improve insulin sensitivity and protect against obesity-induced metabolic syndrome. Due to the presence of their magnetic properties, they show potential against fat reduction processes [42].

12.4.2 Types of Plant-Derived Metal Nanoparticles

(a) *Gold Nanoparticles (AuNPs)*

Gold nanoparticles obtained via green synthesis from plants like *Gynostemma pentaphyllum* and *Smilax glabra* have shown significant anti-obesity activity by modulating fat accumulation and inflammatory cytokines related gene expression and restoring hormone levels of leptin, resistin, adiponectin and the aterogenic and coronary index [43, 44].

(b) *Silver Nanoparticles (AgNPs)*

Artocarpus heterophyllus, *Hippophae rhamniodes* and *Balanites aegyptiaca* derived silver nanoparticles have demonstrated anti-oxidant and regulatory effects on metabolic pathways relevant to obesity management. They also contain various bioactive phytochemicals that regulate metabolic pathways, decrease adipose tissue accumulation, improve glucose tolerance in animal studies, reduce oxidative stress and contribute to modulation of lipid metabolism and anti-obesity effects [45–47].

12.5 Conclusion

Nanoparticles synthesised using a green approach have become a promising class of bio-functional substances to manage and prevent obesity and related disorders. This green approach to make metal oxides from underutilised plants not only makes them sustainable and compatible with the environment but also makes nanoparticles stable with various phytochemical agents present in them. Not only can the metal oxides discussed in this study be used to control obesity, but there are also distinct metal oxides derived from various resources reported by different researchers. Strong antioxidant, anti-inflammatory and metabolic regulatory properties are shown by these NPs which help in lowering oxidative stress, regulating glucose and lipid metabolism, insulin sensitivity and preventing adipogenesis. These metal oxide nanoparticles can regulate oxidative stress, inflammation, glucose homeostasis and lipid metabolism so that they can replace conventional anti-obesity drugs. They have great potential as next generation multi-targeted nanomedicine because of multiple perks as improved bioavailability, tailored activity and less adverse effects as compared to traditional anti-obesity medicines. In conclusion, plant-derived nanoparticles have become an innovative nanomedicine approach to deal with obesity, fat management and other related disorders.

Acknowledgement NA.

Author's Contribution VB, MK—Writing original draft, data collection, conceptualisation; BP—Writing original draft, data interpretation, final formatting; NB—Data interpretation, final formatting.

Competing Interest No.

Ethical Approval Not required.

References

1. Masood B, Moorthy M (2023) Causes of obesity: a review. Clin Med 23(4):284–291. https://doi.org/10.7861/clinmed.2023-0168
2. Omer AE, Shaker G, Safavi-Naeini S (2020) Wearable CSRR-based sensor for monitoring glycemic levels for diabetics. In: 2020 IEEE 20th international conference on bioinformatics and bioengineering (BIBE), Cincinnati, OH, USA, 26-28 October 2020

3. Heymsfield SB, Wadden TA (2017) Mechanisms, pathophysiology, and management of obesity. N Engl J Med 376(3):254–266. https://doi.org/10.1056/NEJMra1514009
4. Li W, Wan H, Yan S et al (2020) Gold nanoparticles synthesized with Poria cocos modulates the anti-obesity parameters in high-fat diet and streptozotocin induced obese diabetes rat model. Arab J Chem 13(7):5966–5977. https://doi.org/10.1016/j.arabjc.2020.04.031
5. Alsenousy AHA, El-Tahan RA, Ghazal NA et al (2022) The anti-obesity potential of superparamagnetic iron oxide nanoparticles against high-fat diet-induced obesity in rats: possible involvement of mitochondrial biogenesis in the adipose tissues. Pharmaceutics 14(10):2134. https://doi.org/10.3390/pharmaceutics14102134
6. Goktas Z, Zu Y, Abbasi M (2020) Recent advances in nanoencapsulation of phytochemicals to combat obesity and its comorbidities. J Agric Food Chem 68(31):8119–8131. https://doi.org/10.1021/acs.jafc.0c00131
7. Rana N, Singh SK, Banu NA et al (2023) The ethnopharmacological properties of green-engineered metallic nanoparticles against metabolic disorders. Medicina (Kaunas) 59(6):1022. https://doi.org/10.3390/medicina59061022
8. Kopelman PG (2000) Obesity as a medical problem. Nature 404(6778):635–643
9. Dubern B, Mosbah H, Pigeyre M (2022) Rare genetic causes of obesity: diagnosis and management in clinical care. Ann Endocrinol 83(1):63–72. https://doi.org/10.1016/j.ando.2021.12.003
10. Beccuti G, Pannain S (2011) Sleep and obesity. Curr Opin Clin Nutr Metab Care 14(4):402–412. https://doi.org/10.1097/MCO.0b013e3283479109
11. Kline CE (2014) The bidirectional relationship between exercise and sleep: implications for exercise adherence and sleep improvement. Am J Lifestyle Med 8(6):375–379. https://doi.org/10.1177/1559827614544437
12. Fava M (2000) Weight gain and antidepressants. J Clin Psychiatry 61(11):37–41
13. Russell-Jones D, Khan R (2007) Insulin-associated weight gain in diabetes–causes, effects and coping strategies. Diabetes Obes Metab 9(6):799–812. https://doi.org/10.1111/j.1463-1326.2006.00686.x
14. Dayabandara M, Hanwella R, Ratnatunga S (2017) Antipsychotic-associated weight gain: management strategies and impact on treatment adherence. Neuropsychiatr Dis Treat 13:2231–2241. https://doi.org/10.2147/NDT.S113099
15. Li MF, Cheung BM (2011) Rise and fall of anti-obesity drugs. World J Diabetes 2(2):19–23. https://doi.org/10.4239/wjd.v2.i2.19
16. Shaikh SM, Desai PV (2024) Magnesium oxide nanoparticles administered orally promote degenerative changes and dysfunctioning in the brain. Indian J Exp Biol 62(04):229–237. https://doi.org/10.56042/ijeb.v62i04.495
17. Wang M, Yang Q, Li Y et al (2025) Therapeutic potential of traditional Chinese medicine and mechanisms for the treatment of type 2 diabetes mellitus. Chin Med 20(1):157
18. Padulosi S, Hoeschle-Zeledon I (2004) Underutilized plant species: what are they? LEISA-LEUSDEN 20:5–6
19. Ponticelli M, Hidalgo-García L, Diez-Echave P et al (2023) Solanum aethiopicum L. from the Basilicata region as a source of specialized metabolites with promising anti-obesity effects: phytochemical characterization and *in vivo* investigation in high fat diet-fed mice. Front Pharmacol 14:1306135. https://doi.org/10.3389/fphar.2023.1306135
20. Generalić Mekinić I, Blažević I, Mudnić I et al (2016) Sea fennel (Crithmum maritimum L.): phytochemical profile, antioxidative, cholinesterase inhibitory and vasodilatory activity. J Food Sci Technol 53(7):3104–3112
21. Alkhudhayri AA (2022) A comparative cytotoxic study against breast cancer cells with nanoparticles and rods shaped structures. J King Saud Univ Sci 34(2):101797. https://doi.org/10.1016/j.jksus.2021.101797
22. Liang YC, Yang MT, Lin CJ et al (2016) Bidens pilosa and its active compound inhibit adipogenesis and lipid accumulation via down-modulation of the C/EBP and PPARγ pathways. Sci Rep 6(1):24285

23. Pareek A, Pant M, Gupta MM et al (2023) *Moringa oleifera*: an updated comprehensive review of its pharmacological activities, ethnomedicinal, phytopharmaceutical formulation, clinical, phytochemical, and toxicological aspects. Int J Mol Sci 24(3):2098. https://doi.org/10.3390/ijms24032098
24. Yang Y, Fukui R, Jia H et al (2021) Amaranth supplementation improves hepatic lipid dysmetabolism and modulates gut microbiota in mice fed a high-fat diet. Foods 10(6):1259. https://doi.org/10.3390/foods10061259
25. Chrigui S, Hadj Taieb S, Jemai H (2023) Anti-obesity and anti-dyslipidemic effects of salicornia arabica decocted extract in tunisian psammomys obesus fed a high-calorie diet. Foods 12(6):1185. https://doi.org/10.3390/foods12061185
26. Ma R, Shi L (2014) Phenylboronic acid-based glucose-responsive polymeric nanoparticles: synthesis and applications in drug delivery. Polym Chem 5(5):1503–1518. https://doi.org/10.1039/C3PY01202F
27. Taha NA, Hussein AA, El-Hak HNG et al (2025) A mini-review of nanoparticle therapeutics targeting oxidative stress and inflammation in diabetes. J Basic Appl Zool 86(1):29
28. Radu CF, Fierascu I, Ortan A et al (2021) Plantago media L.—explored and potential applications of an underutilized plant. Plants 10(2):265. https://doi.org/10.3390/plants10020265
29. Singh J, Dutta T, Kim KH (2018) 'Green'synthesis of metals and their oxide nanoparticles: applications for environmental remediation. J Nanobiotechnol 16(1):84
30. Chopra L (2023) Photo-degradation of dyes and drugs using aloe vera synthesized zinc oxide nanoparticles–a review. Mater Today Proc 72:1613–1617. https://doi.org/10.1016/j.matpr.2022.09.413
31. Bahari N, Hashim N, Abdan K (2023) Role of honey as a bifunctional reducing and capping/stabilizing agent: application for silver and zinc oxide nanoparticles. Nanomaterials (Basel) 13(7):1244. https://doi.org/10.3390/nano13071244
32. Selim YA, Azb MA, Ragab I et al (2020) Green synthesis of zinc oxide nanoparticles using aqueous extract of Deverra tortuosa and their cytotoxic activities. Sci Rep 10(1):3445
33. Vijayakumar S, Mahadevan S, Arulmozhi P et al (2018) Green synthesis of zinc oxide nanoparticles using Atalantia monophylla leaf extracts: characterization and antimicrobial analysis. Mater Sci Semicond Process 82:39–45. https://doi.org/10.1016/j.mssp.2018.03.017
34. Devi R, Gayathri R (2014) Green synthesis of zinc oxide nanoparticles by using Hibiscus rosa-sinensis. Int J Curr Eng Technol 44(44):2444–2446
35. Lin H, Han T, Wang J et al (2024) Screening and identification of a strain with protease and Phytase activities and its application in soybean meal fermentation. Appl Biochem Biotechnol 196(2):790–803
36. Venkatappa MM, Udagani C, Hanumegowda SM et al (2022) Effect of biofunctional green synthesized MgO-nanoparticles on oxidative-stress-induced tissue damage and thrombosis. Molecules 27(16):5162. https://doi.org/10.3390/molecules27165162
37. Vergheese M, Vishal Sk, Mary Vergheese C (2018) Green synthesis of magnesium oxide nanoparticles using Trigonella foenum-graecum leaf extract and its antibacterial activity. J Pharmacogn Phytochem 7(3):1193–1200
38. Ali S, Sudha KG, Thirumalaivasan N (2023) Green synthesis of magnesium oxide nanoparticles by using abrus precatorius bark extract and their photocatalytic, antioxidant, antibacterial, and cytotoxicity activities. Bioengineering 10(3):302. https://doi.org/10.3390/bioengineering10030302
39. Jeevanandam J, Chan YS, Danquah MK (2020) Cytotoxicity and insulin resistance reversal ability of biofunctional phytosynthesized MgO nanoparticles. 3 Biotech 10(11):489
40. Kanagasubbulakshmi S, Kadirvelu K (2017) Green synthesis of iron oxide nanoparticles using Lagenaria siceraria and evaluation of its antimicrobial activity. Def Life Sci J 2(4):422–427. https://doi.org/10.14429/dlsj.2.12277
41. Karpagavinayagam P, Vedhi C (2019) Green synthesis of iron oxide nanoparticles using Avicennia marina flower extract. Vacuum 160:286–292. https://doi.org/10.1016/j.vacuum.2018.11.043

42. Yusefi M, Shameli K, Su Yee O et al (2021) Green synthesis of Fe3O4 nanoparticles stabilized by a Garcinia mangostana fruit peel extract for hyperthermia and anticancer activities. Int J Nanomed:2515–2532. https://doi.org/10.2147/IJN.S284134
43. Ansari S, Bari A, Ullah R et al (2019) Gold nanoparticles synthesized with Smilax glabra rhizome modulates the anti-obesity parameters in high-fat diet and streptozotocin induced obese diabetes rat model. J Photochem Photobiol B 201:111643. https://doi.org/10.1016/j.jphotobiol.2019.111643
44. Sharma VK, Prateeksha, Gupta SC et al (2022) *Cinnamomum verum*-derived bioactives-functionalized gold nanoparticles for prevention of obesity through gut microbiota reshaping. Mater Today Bio 13:100204. https://doi.org/10.1016/j.mtbio.2022.100204
45. Bhardwaj M, Yadav P, Yadav M et al (2024) Phytochemical screening and antidiabetic efficacy of Balanites aegyptiaca seed extract and their silver nanoparticles on muscle and pancreatic cell lines. ACS Omega 9(21):22660–22676. https://doi.org/10.1021/acsomega.4c00327
46. Rana N, Banu AN, Kumar B et al (2024) Phytofabrication, characterization of silver nanoparticles using Hippophae rhamnoides berries extract and their biological activities. Front Microbiol 15:1399937. https://doi.org/10.3389/fmicb.2024.1399937
47. Bagya Lakshmi J, Vibisha K, Vignesh M et al (2025) Green synthesized silver nanoparticles of Artocarpus Heterophyllus peel extract and its evaluation for anti-inflammatory activity. J Clin Res Notes 6(2). https://doi.org/10.31579/2690-8816/163
48. Roghini R, Muthulakshmi M, Selvi K et al (2025) Green synthesis of silver nanoparticles using Persea Americana fruit extract prevents adipogenesis of 3t3-L1 cells via down regulating Ppar-γ. J Neonatal Surg 14:188–197. https://doi.org/10.52783/jns.v14.2701
49. Chen H, Ng JPM, Tan Y et al (2018) Gold nanoparticles improve metabolic profile of mice fed a high-fat diet. J Nanobiotechnol 16(1):11. https://doi.org/10.1186/s12951-018-0338-1
50. You W, Ahn JC, Boopathi V (2021) Enhanced antiobesity efficacy of tryptophan using the nanoformulation of Dendropanax morbifera extract mediated with ZnO nanoparticle. Materials 14(4):824. https://doi.org/10.3390/ma14040824
51. Kumar M, Kaushik D, Kumar A et al (2024) A sustainable approach to prepare green synthesis of copper nanoparticles of Bauhinia variegata & Saussurea lappa: unveiling in-vitro anti-obesity applications. Heliyon 10(8):e29433. https://doi.org/10.1016/j.heliyon.2024.e29433
52. Alshammari GM, Yagoub AEA, Subash-Babu P et al (2021) Inhibition of lipid accumulation and adipokine levels in maturing adipocytes by Bauhinia rufescens (lam.) stem bark extract loaded titanium oxide nanoparticles. Molecules 26(23):7238. https://doi.org/10.3390/molecules26237238
53. Yagoub AEA, Alshammari GM, Subash-Babu P et al (2021) Synthesis of Ziziphus spina-christi (Jujube) root methanol extract loaded functionalized silver nanoparticle (ZS-Ag-NPs); physiochemical characterization and effect of ZS-Ag-NPs on adipocyte maturation, adipokine and vascular smooth muscle cell interaction. Nanomaterials 11(10):2563. https://doi.org/10.3390/nano11102563

Glossary

Angiogenesis The process of developing new blood vessels from previously formed blood vessels that is important for normal growth and development, as well as for healing from injuries. This process may also play a role in diseases, such as tumors and chronic inflammatory conditions.

Antimicrobial Resistance (AMR) A condition where microorganisms—including bacteria, fungi, viruses, and parasites—can survive and/or grow in the presence of antimicrobial agents that were previously effective at killing or inhibiting them. It usually develops from genetic mutations or the acquisition of resistance genes and creates a significant challenge to public health because it reduces the efficacy of standard therapies and increases the likelihood of prolonged infection.

Bimetallic nanoparticles These types of nanoparticles consist of two different metals formed into one nano-dimensional structure (e.g., combined to create an alloy). It can exist as either a core-shell structure, or as a heterostructure NPs, in which two metals come together to form a single NP. Bimetallic nanoparticles can be made from a variety of different metals. Three examples of bimetallic nanoparticles include: Ag-Cu, Zn-Cu, and Ag-ZnO.

Bioreductant A biomolecule or bioproduct that donates electron to reduction of either metal ions. This would be the phytochemicals, enzymes, and metabolites produced from plants, fungi, microbes, etc., used widely in the environment-friendly synthesis of nanoparticles or any other reduced products.

Hypoglycemia and Hyperglycemia Hypoglycemia is a result of low levels of blood glucose. Symptoms associated with this condition may include dizziness, confusion, or loss of consciousness, while hyperglycemia is the result of high levels of blood glucose, usually seen in patients with diabetes mellitus, and can cause both immediate and long-term vascular and metabolic problems if not treated properly.

Maceration It is a technique that involves the use of a solvent to soften and break down plant or biological tissue at room temperature for the extraction of the

© The Editor(s) (if applicable) and The Author(s), under exclusive license to Springer Nature Switzerland AG 2026

E. Nepovimova et al. (eds.), *Underutilized Plants Based-Nanoparticles: Herbal Remedy to Treat Human Diseases*, https://doi.org/10.1007/978-3-032-20384-7

active compounds through diffusion into the solvent. This is a widely used technique in many areas including phytochemistry, herbal medicine, and research on natural products.

Metallic Nanoparticles (NPs) They are defined as nanoparticles consisting of a single metallic element, again from 1 to 100 nm in size. A very important aspect of these NPs is that their unusual atomic size confers specific optical, catalytic, and biological properties that make them attractive for use in many applications. Examples of these types of NPs include gold (AuNPs), silver (AgNPs), copper (CuNPs), zinc (ZnNPs), and iron-based (FeNPs) nanoparticles which are being studied in a large number of applications ranging from industrial to environmental to biomedical.

Nanoencapsulation The use of nanocarriers (1–1000 nm) to encapsulate (and coat) active compounds, such as bioactive molecules, drugs, nutrients, and antioxidant, has gained popularity in various industries due to its ability to enhance the stability, solubility, bioavailability, and controlled release characteristics of these compounds. These properties make the nanocarrier technology applicable in pharmaceuticals, food, cosmetics, and biomedical applications.

Nanoparticles (NPs) It refers to a certain type of material that is defined by their physical size typically measuring from 1 to 100 nm in physical measurement. The term "nanoparticle" has been defined to describe those unique properties of the nanoparticle material that occur as a direct result of the nanoscale size of the material compared to bulk materials.

Underutilized Plants There are numerous plant species that have been historically utilized (e.g., harvested) for many purposes (e.g., consumption of food, feed for animals, and for traditional medicine) but still have the potential to provide additional benefits to humanity in terms of socio-economic value. Such species may also be referred to as "under-utilized," "minor," "underdeveloped," "underutilized/underexploited," "orphan," "promising," or "alternative plant" species.

Valorization It is the process of adding value to material, resource, or by-product through conversion into more useful, marketable, or higher value products. Valorization is used most commonly in the fields of agriculture, biotechnology, environmental science, and industrial processes as a means of improving resource efficiency and sustainability and reducing waste.

MIX
Papier aus verantwortungsvollen Quellen
Paper from responsible sources
FSC® C105338

If you have any concerns about our products,
you can contact us on
ProductSafety@springernature.com

In case Publisher is established outside the EU,
the EU authorized representative is:
Springer Nature Customer Service Center GmbH
Europaplatz 3, 69115 Heidelberg, Germany

Printed by Libri Plureos GmbH
in Hamburg, Germany